The BOURBON TASTING *Notebook*

THIRD EDITION

SUSAN REIGLER

MICHAEL VEACH

Acclaim Press
MORLEY, MISSOURI

Acclaim Press
— Your Next Great Book —

P.O. Box 238
Morley, MO 63767
(573) 472-9800
www.acclaimpress.com

Book & Cover Design: M. Frene Melton

ISBN: 978-1-956027-59-4 | 1-956027-59-9
Library of Congress Control Number: 2023938401

Third Edition, First Printing: 2024
Printed in the United States of America
10 9 8 7 6 5 4 3 2 1

This publication was produced using available information.
The Publisher regrets it cannot assume responsibility for errors or omissions.

Contents

For Joanna
and
For My BFF

Preface

The tide of bourbon has risen to flood stage. This third edition of *The Bourbon Tasting Notebook* contains more than double the number of bourbons than the first edition, so you will find more than 500 bourbons to try, take notes about, and score. Most of the added bourbons come from small and medium-sized distillers and independent bottlers. But many of the major distilleries continue to release new bourbons, as well as new expressions of many of their brands. While Kentucky still produces the lion's share of bourbon, this book contains bourbon from thirty-eight states and the District of Columbia.

A few of the bottles we have in the book are no longer being made, or were limited releases. We have kept them (and noted when this is the case) because you may come across them in one of the larger bourbon bars around the country. Tip: Always ask if a bar or restaurant has a private selection bottling. It will be intriguing to compare it to the regular expression of the brand.

Happy tasting, readers!

Cheers,
Susan & Michael

Nuts and Bolts

How to Use This Book

With a record amount of bourbon whiskey aging in warehouses and visits to Kentucky's bourbon distilleries topping 3 million annually, it is obvious that bourbon is enjoying unprecedented popularity. Just as people serious about wine or craft beer or Scotch like to keep track of what they have tasted and what labels are their favorites, we know that bourbon lovers are the same.

So, our purpose in writing this tasting notebook was to give you, the bourbon enthusiast, a handy logbook for keeping track of your sampling. Unlike the authors of many other beverage tasting guides, we have not attempted to say which bourbons are best or to assign ratings. We are leaving that up to you!

Taste is a highly subjective thing. For example, while we very often liked the same bourbons, we almost as often liked them for different reasons, detecting slightly different flavors, or similar flavors at different stages of a tasting. (Mike may have picked up a particular spicy or fruity note on the nose that Susan didn't experience until the finish, and vice versa.)

We also want you to know that we tasted no more than five or six bourbons per session, so our palates would stay fresh. The following information is given for each bourbon:

Proof: Ranging from 80 proof to barrel proof.

Age: When stated on the label, we included. If not, you'll see NAS for No Age Statement. Some distilleries will state on their websites or on tours that their whiskies are aged over a certain time span and we have included that information when we have been told.

Type: Most are straight bourbons meaning they have been aged at least two years and what's in the bottle is all bourbon. A handful of products were blended, that is, neutral grain spirit was added to the whiskey. These are designated "bourbon a blend."

Style: When the bourbon has a particular style, such as extra aged, barrel proof, or wheated, that style is noted. Otherwise, style is "Traditional."

Mash Bill: Some distilleries share their grain recipes. We give them where we know them. Others will say they use one or several and we describe this in general terms in the introductions. For example, Wild Turkey uses the same mash bill for all of its bourbons and simply states it is "high in rye."

Color: From lightest to darkest: Pale Straw, Light Straw, Straw, Dark Straw, Light Amber, Amber, Dark Amber, Bronze. (Straw has yellow hues. Ambers contain reddish ones.) Often, but not always, the longer the bourbon has aged, the darker the color. Higher proof bourbons may be darker, too.

Price: Price scale is based on what you would pay for the bourbon in Kentucky or, in the case of a bourbon not made in Kentucky, its state of origin. In any case, the price is relative and subject to change. Prices may be higher in other markets.

$ = Under $25
$$ = $25–$49.99
$$$ = $50–$99.99
$$$$ = $100 and above

Nose: We let the bourbon warm in a Glencairn tasting glass before nosing to detect characteristic aromas. To buy your own Glencairn glasses, go to a local liquor store or www.whiskyglass.com. You can also use any tulip shaped glass, such as a white wine or sherry glass, or a snifter.

Taste: We usually sipped without water, but we sometimes added a few drops of water to higher proof bourbons, especially those bottled at barrel proof.

Finish: The flavors revealed on the palate after the bourbon has been swallowed.

Notes: Our general comments about the whiskey, which may include overall impressions, suggestions for enjoyment, and other information.

We have not assigned numerical ratings to the bourbons. We are leaving that to you, since only you know which bourbons you like and for what reasons. Space is provided for your notes and ratings for each.

We have listed the bourbons by distillery and alphabetically within each distillery listing. A few special bottlings, such as annual limited releases, were tasted. Bear in mind that these can vary from year to year. We have not included whiskies to which flavoring has been added, such as Red Stag. By law, these are no longer bourbons. Major Kentucky distilleries, since they make approximately 95% of bourbons, are listed first. Then, there is an alphabetical listing of expressions produced by smaller and medium sized distilleries in Kentucky and other states, as well as by independent bottlers, the brands that buy whiskey from other distilleries. Many of them rebarrel that bourbon to create distinctive flavor profiles.

All names are indexed alphabetically and to help you make comparisons, we have indexed the bourbons by proof, price, and style. This edition adds an index by state, too.

Bourbon Defined

All bourbon is whiskey, but not all whiskey is bourbon. To understand bourbon's place in the world of whiskey, you must first understand "what is whiskey?" The word "whisky" is an Anglicized form of the Gaelic word "uisge beatha" (pronounced WEEZ-ga-bochh) or "water of life"; in Latin, "Aqua Vita", is the term used for all distilled spirits five hundred years ago. "Uisge" came to be the spirit distilled from the beers of Scotland and Ireland, as opposed to the brandies from the continental distilleries.

Today, the modern definition of whiskey is very simple — It is a distilled spirit made from a fermented cereal grain or a "beer". This "beer" is distilled at no higher than 160 proof and the spirit is placed in an oak container to age. Once removed from the oak, a whiskey must be bottled at no less than 80 proof. These simple rules define whiskey as a category. Different styles of whisky set additional rules such as age or geographical requirements. As long as these basic rules are not broken, then it is a "whiskey".

Bourbon is a whiskey with refined rules setting it apart from other whiskey styles. These rules evolved over the past two hundred years, creating the product we know today as "bourbon". The bourbon rules are the whiskey rules but narrowed and refined as follows:

1. Bourbon has to be at least 51% corn. This is a minimum amount of corn and there is no upper limit for the amount of corn. Bourbon can be made from 100% corn, if that is what the distillers wants.
2. Bourbon cannot be distilled at higher than 160 proof. This allows more flavors from the grain and yeast to come through in the final distillate.
3. Bourbon has to go into a brand new, charred oak container at no more than 125 proof, and only pure water can be added to adjust the proof.
4. Bourbon has to be bottled at least at 80 proof.
5. Bourbon has to be a product of the United States.

There is no age requirement for bourbon. As soon as the distillate enters the barrel, it is "Bourbon". There are a few non-bourbon rules that must be adhered to in order for it to be called "Straight Bourbon".

1. To be called, "Straight" the whiskey has to be aged for two years in a new oak container.
2. Nothing other than pure water can be added to the whiskey to adjust the proof. No artificial coloring or flavoring is allowed.
3. If the distiller wants to bottle a bourbon without an age statement, then the whiskey inside the bottle has to be at least four years old.

These rules are the rules for making bourbon whiskey. They are more narrowly defined than the rules for "Whiskey", but they still have plenty room to allow distillers to create their own unique product.

There are six ways that the flavor can be varied using these rules. The distilleries use these variations to create their own flavor profiles for their brands. Here are the flavor variations:

1. **Grain**: The grain is a major source of flavor for bourbon. Bourbon has to be at least 51% corn, but the amount of corn is a factor in defining the flavor of the final product. Traditionally, bourbon uses a percentage

of rye to help flavor the whiskey, and the amount of rye in the mash bill will determine spice and fruit notes in the bourbon. Some brands such as Maker's Mark, Weller and Old Fitzgerald use wheat instead of rye in their mash bill, giving the bourbon a softer profile with more nuttiness in the flavor. Barley malt was needed to make bourbon in the days before artificial enzymes in order to convert starches into sugars for fermentation. If a distiller is using less than 10% barley malt, they are also using artificial enzymes for this product. Barley malt gives the bourbon some sweet nutty flavors in the distillate.

2. **Water**: The source of water has an impact on the flavor of bourbon. Some distilleries use spring water from deep limestone wells that are full of minerals that will add flavor to the spirits produced. Others use the local city water that has been treated through a process of reverse osmosis (RO water), removing all minerals and leaving just pure water. The important part of the Kentucky water is that it is iron free water, as iron adds a bad flavor to distilled spirits.
3. **Fermentation**: There are thousands of different variations of yeast, and each strain produce different flavors of alcohol. The Four Roses distillery has five different strains of yeast: Light Fruit, Heavy Fruit, Spicy, Herbal, and Floral. Each strain produces these flavors during fermentation. Brown-Forman has yeast strains for Early Times, Old Forester and Jack Daniels. The yeast they use for Woodford Reserve started as Old Forester yeast, but mutated to create the flavors found in Woodford. Yeast plays an important part of creating the flavor profiles in bourbon, as each distillery uses different strains of yeast and often in different ways. Some distillers use jug yeast grown at the distillery, while others use dried yeast purchased from a yeast producing company. Many distilleries allow for a three day fermentation while others will go four, five or even six days of fermentation. These practices all affect the flavor of the final spirit.

4. **Distillation**: This includes the type of stills used to make the distillation. Most distilleries use a column still with a pot still doubler. Some use a Thumper. Woodford Reserve uses pot still and triple distills the spirit. Many craft distilleries are using small hybrid stills that act as a pot still with a small rectifying column attached. The materials used to make the stills are important. Stills may be all copper, or they may be stainless steel with copper heads, or they may be 100% stainless steel with copper wool placed inside the column. Besides the equipment used, the distillation proof of the spirit is a factor in the flavor. The higher the distillation proof, the less flavor from the first three factors are carried into the spirit produced.
5. **Maturation**: This includes many factors such as the barrel, warehouse and the number of years aged. With the barrel, the distiller has to consider the level of char on the inside of the barrel and whether to "toast" the wood deeply before charring. The warehouse used also has an impact on flavor. Is it constructed of brick or stone or is it an iron clad warehouse with little insulation value to its walls? Is it heat cycled in the winter? Is it a warehouse with barrel racks or are the barrels being stored on pallets? Is it a single story warehouse or multi-storied? If it is multi-storied, what floor of the warehouse is the barrel stored? Is the barrel stationary once it is entered into the warehouse or does the distillery rotate barrels from level to level? Is the warehouse on a hillside in the country or in the heart of town? Finally, the distiller must decide how long to leave the barrel of bourbon in the warehouse before bottling the final product.
6. **Bottling**: Not only is age important in the flavor of the final product, but so are the bottling practices. The distiller has to decide the final proof in which to bottle the product. The type of filtration will have an effect on the flavor, even if the choice is not to filter at all. Is the product going to be a bonded

product, with all of the whiskey coming from the same distillery and made in the same season, or is it going to be a mixture of different ages and possibly even bourbon from a different distillery? Four Roses uses ten different bourbons that they make at the same distillery to create their Four Roses Yellow Label. Is the final product a small batch product made from a limited number of barrels or even a single barrel product? These factors all help determine the final flavor profile of the bourbon that is bottled and sold to the consumer.on that is bottled and sold to the consumer.

Even though all bourbon distilleries follow the bourbon regulations, each distiller manipulates these six flavor variables to create their own unique flavor profile for their brands. Many times, a distillery will use bourbon made with the first four variables the same but will use maturation and bottling to create multiple brands of bourbon, each with its own flavor profile.

Experiment with glassware by taking several different styles of glass and pour the same bourbon in each and then nose the bourbon in each glass. Make a note as to how the aromas are different in each glass.

Tasting Bourbon

Tasting bourbon is not a complex process, but the taster should have some understanding of what is happening so that he or she can determine the "why I like this" instead of simply saying "this is good". The first step in this process is to understand glassware, since the type of glass will change the tasting experience. For this book, the Glencairn Whisky Tasting Glass was used to taste the bourbons. It was chosen because it is a very good design for tasting whiskey and it is available for purchase at a reasonable price in most distillery gift shops, in many liquor stores, or from online sources, allowing you to acquire the glass for yourself. What makes the glass a good choice is the fact that the inward taper of the glass funnels aromas to the nose. If a Glencairn glass is not available, then a brandy snifter or tapered wine glass will work.

When tasting bourbon, you want to use your senses of sight, smell and taste. Each sense will help you to better understand what you are drinking. We will start with the sense of sight and color.

1. **Color**: Color will tell you something about the age and proof of the bourbon. As the spirit ages in a barrel, it gets darker in color as it picks up wood tannins from the charred oak. This color will vary from a light straw yellow to a dark amber red with age. Bourbon has no artificial caramel coloring added, so all of the coloring comes from aging in every bourbon, no exceptions. Adding water to adjust the proof will then lighten the color of the bourbon, so the whiskey at barrel proof will be much darker than the same whiskey at 80 proof. Not only is the water lightening the color, but also the whiskey is more heavily filtered at lower proof, removing more color in the process.
2. **Nose**: The aromas found in a bourbon will tell the taster much about what to expect to find in the taste. There are several different categories of aromas that are detected in bourbon. Not every bourbon will have aromas in every category, because every bourbon is different. Here are the categories:
    a. Candy Shop: These are the sweet candy-like aromas found in bourbons and usually created in aging whiskey. This category includes caramel, vanilla, toffee, maple syrup, chocolate – milk and dark varieties and pralines.
    b. Wood Shop: These are aromas from the barrel, but also the nut aromas that often come from the grains. This category includes oak, cedar, charcoal smoke, pecan, hazelnut, walnut, hickory nut, and sawdust or fresh cut wood.
    c. Fruit Stand: Flavors that often come from the yeast, but grain recipe can have some contribution to

these aromas. They include apple (green and ripe), pear, cherry, berry (raspberry, blueberry, strawberry, banana, apricot, peach, date), citrus (orange, lemon, grapefruit, lime), and grape or wine aromas.

d. Flower Shop: Floral aromas often created in fermentation from the yeast. These include rose, lilac, honeysuckle and orange blossoms.

e. Herbal Shop: Aromas most often created in fermentation, but also in the aging process. This category includes mint, tobacco, licorice or anise and other herbal aromas.

f. Spice Rack: These flavors come from the grain, fermentation and aging of bourbon. They include clove, nutmeg, cinnamon, allspice, and pepper (black and white).

Adding water experiment: Take a straw or dropper and add water to a glass of bourbon three drops at a time and note how the water changes the aromas and flavor of the bourbon at each stage.

Bad aromas can be found in bourbon, but it is hoped the distiller detected them in samples and never let such whiskey into the bottle. Still, sometimes the cork will be bad, giving a bottle of bourbon a bad musty aroma and taste. Bad flavors include must, mold, dirt, cabbage or skunk (caused by sulfur) and acetone or chemical aromas.

3. **Taste**: There are many different ideas about how you must taste bourbon. Some will say to open your mouth after swallowing, while others will say to chew on the bourbon. The correct way to taste is whatever is most natural for the taster. The important thing to remember is to let the bourbon to flow over the tongue, covering taste buds in every region of the mouth. That is why

swallowing the bourbon is important, because 40% of the taste buds are locate in the esophagus and that is where a lot of the finish takes place. The flavors found can be split into the same six categories that are found in the nose, but they may be very different from what was detected on the nose. Adding a little water or letting the bourbon breath can often change the taste of a bourbon.

4. **Finish**: The Finish is the aftertaste left after swallowing the bourbon. This can be very long or very short, depending upon the bourbon. It can be very sweet with candy, fruit or sweet spice flavors or very dry with wood tannins and peppery spices. It is not uncommon for the finish to start one way and change to another before it is over.

Pour a glass of bourbon and taste it immediately. Then let the bourbon sit for five minute intervals between tastes and note how the aromas and flavors change.

Hosting a Bourbon Tasting for friends and family can be a very entertaining event. It is a very easy event to put together, and here are some steps to take to make the process easy.

1. Choose a theme for the tasting. Themes make the tasting a learning experience as well as a fun night of enjoying bourbon. There are many themes that can be used for a tasting. Some favorites include: a) Traditional bourbon made with rye and "wheated" bourbon; b) Same brand, different age; c) Same brand, different proof; d) Different brand, same distillery; e) Bourbon, Rye, Tennessee whiskey comparison. There are many more themes that can be used for a tasting, so feel free to experiment with different themes.

2. Choose glassware. The glassware should be the same for every taster so that they are getting the same experience with the bourbon.
3. Make sure the tasting area is clear of interference from strong smelling candles, tobacco smoke, cooking odors and other aromas that can disrupt the experience. Some even prefer to keep other outside distractions to a minimum and turn off televisions and music in the area.
4. Make sure that there is plenty of water available for those who wish to add a little water to the bourbon and to cleanse the palate between tastings.
5. Follow the steps provided above and make note of the tasting experience by either writing it down in your notebook or at least discussing the tastes and aromas amongst the group.

The tasting can also involve food. Traditionally, only a bland food such as a cracker is used to cleanse the palate between products. However, Ouita Michel, the former chef-in-residence at Woodford Reserve Distillery has developed a flavor wheel of foods to help people understand how bourbon can be enjoyed with food. She takes a small plate and places sample bites of Parmesan cheese, dried cranberries or cherries, roasted hazelnuts or pecans, a fresh orange slice, a piece of dark chocolate and a spoon of sorghum. She has varied these foods from time to time and has used olives, orange zest, malt syrup, and other foods on the wheel as well. The point of the wheel is to take a sip of bourbon before eating anything on the plate. This establishes a flavor profile for the bourbon in the taster's mind. The taster then picks up a sample of the food and eats it, followed by another sip of the bourbon. The changes of flavor in the bourbon and or the food are then noted. This helps her as a chef to decide what foods are best paired with bourbon, as well as how bourbon can be used in her recipes. This type of tasting can be used with any of the themes listed above or even become a tasting theme in itself.

Major Kentucky Distilleries

Angel's Envy

www.angelsenvy.com

Bourbon Distilleries:

- Angel's Envy Distillery

The company was founded in 2010 by Lincoln Henderson, who had retired as the master distiller at Brown-Forman, and his son Wes. While at Brown-Forman, Henderson had been responsible for creating the premium Woodford Reserve brand. He had also done some experimenting with finishing bourbon in barrels that had stored other sprits and so it was determined that Angel's Envy would receive secondary barrel aging in ruby port barrels.

The name is a play on "the angel's share" the portion of whiskey that evaporates into the atmosphere during ageing. Henderson was obviously confident that the angel's would be disappointed that they didn't get access to what hadn't escaped from the barrel, but instead wound up in the bottles.

Angel's Envy

Proof: 86.6
Age: NAS
Type: Straight
Style: Special Finish, Small Batch
Mash Bill: 72% corn, 18% rye, 10% malted barley
Color: Straw
Price: $$

Mike

Nose: Vanilla and fruit with some caramel and corn and hint of fine tobacco.
Taste: Sweet fruit and caramel with tobacco and oak.
Finish: Long and sweet with the port wine coming out at the end. Some oak adds balance.

Susan

Nose: Very fruity, with good vanilla and caramel notes.
Taste: Lovely vanilla on the tongue with dark fruit and light spices and a wine-like note.
Finish: Long and warm.

Notes

Batch No. FA14, Bottle No. 1. Finished by aging in used port wine casks. Sourced from Kentucky distilleries and selected by the late Lincoln Henderson, retired Master Distiller from Brown-Forman who created Woodford Reserve Distributed by Louisville Distilling Company, Crestwood, Kentucky. The company opened its own distillery on Main Street in Louisville in 2016 and is now producing whiskey there.

My Score

Notes: ____________________________________

____

____

Appendearance (1-5) ___ Nose (1-25) ___
Taste (1-25) ___ Finish (1-25) ___
Complexity (1-10) ___ Overall Impression (1-10) ___

SCORE ___

Angel's Envy Cask Strength Bottled 2012

Proof: 123.7
Age: NAS
Type: Straight
Style: Special Finish, Small Batch
Mash Bill: 72% corn, 18% rye, 10% malted barley
Color: Amber
Price: $$$

Mike

Nose: Rich caramel – almost molasses – with berries and grapes and a hint of tobacco. Water. weakens the nose, but brings out the fruit.
Taste: Rich fruit and caramel, a bit if tobacco and oak. Water sweetens it (more fruit than caramel) and amplifies the tobacco.
Finish: Long and dry with oak tannins. Water does not change the finish.

Susan

Nose: Rich caramel, pecans, cocoa.
Taste: Lots of oak at barrel proof. Water brings out licorice, pecan, toffee, and dried fruit.
Finish: Long, warm, and nutty. Elegant.

Notes

Save this for a snifter in an armchair by the fireplace. There is a limited release of this expression each year. Cask strength varies. For example, the 2014 for 119.3 proof.

My Score

Notes: ______________________________

____

____

Appearance (1-5) ___ Nose (1-25) ___
Taste (1-25) ___ Finish (1-25) ___
Complexity (1-10) ___ Overall Impression (1-10) ___
SCORE ___

Angel's Envy Oloroso Sherry Cask Finish

Proof: 100
Age: At least 4 years
Type: Straight
Style: Finished
Mash Bill: 72% corn, 18% rye, 10% malted barley
Color: Straw
Price: $$$$

Mike

Nose: Caramel, chocolate, almonds, and berries with a hint of oak.
Taste: Lots of wood tannins, caramel, berries, and pepper spice.
Finish: Long and dry with oak and pepper, but a little fruit lingering in the background.

Susan

Nose: Stewed prunes with sweet spices and notes of caramel and chocolate.
Taste: Sweet floral flavor. Orange blossom? Also, caramel, vanilla, faint whiffs of allspice and nutmeg.
Finish: Rather short. Dry without bitterness.

Notes

Limited edition, but you'll be able to find it in well-stocked bourbon bars. Look for other limited expressions including finished in madeira and ice cider casks.

My Score

Notes: ______________________________

____

____

Appearance (1-5) ____ Nose (1-25) ____
Taste (1-25) ____ Finish (1-25) ____
Complexity (1-10) ____ Overall Impression (1-10) ____
SCORE ____

Angel's Envy Additional Releases

Name: ______________________________

Distillery: ______________________________

Proof: ____________ **Mash Bill:** ____________
Age: ____________ ____________
Type: ____________ **Color:** ____________
Style: ____________ **Price:** ____________

My Score

Notes: ______________________________

____

____

Appearance (1-5) ____ Nose (1-25) ____
Taste (1-25) ____ Finish (1-25) ____
Complexity (1-10) ____ Overall Impression (1-10) ____
SCORE ____

Name: ______________________________

Distillery: ______________________________

Proof: ____________ **Mash Bill:** ____________
Age: ____________ ____________
Type: ____________ **Color:** ____________
Style: ____________ **Price:** ____________

My Score

Notes: ______________________________

____

____

Appearance (1-5) ____ Nose (1-25) ____
Taste (1-25) ____ Finish (1-25) ____
Complexity (1-10) ____ Overall Impression (1-10) ____
SCORE ____

Bardstown Bourbon Company

www.bardtownbourbon.com

Bourbon Distilleries:

- Bardstown Bourbon Company, Bardstown, Ky.

While it has only been around since 2016, Bardstown Bourbon Company has rapidly grown into a major distillery. Not only does it release its own brands, but it contract distills for a slew of independent bottlers. (And, at least at one time, made juice for some major brands that had run out of capacity. But they are not saying who.) The site also features a restaurant and cocktail bar and a large library of vintage whiskeys of all brands. Tastes are offered for all of them, the rearest of which for hefty prices.

Collaboration American Brandy Finish

Proof: 113
Age: 10 years
Type: Straight
Style: Finished
Mash Bill: 75% corn, 21% wheat, 4% malted barley
Color: Light Amber
Price: $$$

Mike

Nose: Very fruity. Raisins and apples with a hint of oak, vanilla, and spice.
Taste: Fruit forward. Raisins and apples with pepper spice, oak, and caramel.
Finish: Long and dry with oak.

Susan

Nose: Grapes, caramel, almonds.
Taste: Caramel and brown sugar. Water amplifies the grapes.
Finish: Long with spicy oak.

Notes

One of BBC's Collaboration Series releases. The brandy barrels used were from Coper & Kings Distillery in Louisville.

My Score

Notes: ______________________________

____

____

Appearance (1-5) ___ Nose (1-25) ___
Taste (1-25) ___ Finish (1-25) ___
Complexity (1-10) ___ Overall Impression (1-10) ___
SCORE ___

Collaboration Copper & Kings Double Mistelle Finish

Proof:	94
Age:	10 years
Type:	Straight
Style:	Finished
Mash Bill:	75% corn, 21% rye, 4% malted barley
Color:	Straw
Price:	$$$

Mike

Nose: Very fruity – raisins and raspberries dominate the nose. Hints of vanilla and spice.

Taste: Raisins and raspberries joined by a hint of oak and spice.

Finish: Long and fruity.

Susan

Nose: Fruity wine notes with some vanilla and cashews.

Taste: Very sweet. Sugar coated nuts with ripe fruit – grapes.

Finish: Long and sweet with a spike of vanilla at the very end.

Notes

A perfect bourbon for brandy drinkers.

My Score

Notes: ____________________________________

____

____

Appearance (1-5)	____	Nose (1-25)	____
Taste (1-25)	____	Finish (1-25)	____
Complexity (1-10)	____	Overall Impression (1-10)	____

SCORE ____

Discovery Series #1

Proof: 121.1
Age: 5, 10, 11, and 13-year blend
Type: Straight
Style: Blend, Cask Strength
Mash Bill: Complicated. See website.
Color: Light Amber
Price: $$$$

Mike

Nose: Lots of oak, caramel, ripe apples, and pepper.
Taste: Oak forward with caramel, pepper, a little apple, and leather.
Finish: Long, dry, and spicy.

Susan

Nose: Caramel, crème brulee, milk chocolate, and brown sugar.
Taste: Caramel, cinnamon, apple peel, and some vanilla.
Finish: Dries to oak.

Notes

Water releases the apples. There have been several more releases in the series. See website for a complete list.

My Score

Notes: ____________________________________

____

____

Appearance (1-5) ____ Nose (1-25) ____
Taste (1-25) ____ Finish (1-25) ____
Complexity (1-10) ____ Overall Impression (1-10) ____

SCORE ____

Fusion Series #1

Proof: 98.9
Age: 2 years, 1 month (See notes)
Type: Straight
Style: Four Grain
Mash Bill: See notes.
Color: Light Straw
Price: $$$

Mike

Nose: Toasted marshmallows, a little fruit, spice, and oak.
Taste: Vanilla, apples, baking spices, and oak.
Finish: Long, dry with oak and spices.

Susan

Nose: Vanilla, apple, banana, and some nutmeg.
Taste: Rich vanilla followed by the fruit with a sprinkling of spice.
Finish: Sweetness quickly dries to peppery oak.

Notes

A blend of 40% sourced (Kentucky) 11 year, 7-month, 18% BBC's 2 year, 3-month wheated bourbon, and BBC's 2 year, 1-month, rye bourbon.

My Score

Notes: ____

____

____

Appearance (1-5) ____ Nose (1-25) ____
Taste (1-25) ____ Finish (1-25) ____
Complexity (1-10) ____ Overall Impression (1-10) ____
SCORE ____

Origin Series Bottled-in-Bond Kentucky Straight Bourbon Whiskey

Proof: 100
Age: 6 years
Type: Straight
Style: Wheated, Bonded
Mash Bill: 68% corn, 20% wheat, 12% malted barley
Color: Straw
Price: $$$

Mike

Nose: Vanilla, marshmallows, apples, white pepper, oak.
Taste: Vanilla, ripe apple, white pepper, oak.
Finish: Medium long with oak, pepper, and lingering apple.

Susan

Nose: Cream soda, apple peel, some sweet cherries, and light spice.
Taste: Vanilla, dates, light pepper, and oak.
Finish: Long and tingly.

Notes

The Origin Series is the first collection of releases from Bardstown Bourbon Company make exclusively from whiskey distilled at its facility and not blended with other sourced whiskeys.

My Score

Notes: ______________________________

____

____

Appearance (1-5) ____ Nose (1-25) ____
Taste (1-25) ____ Finish (1-25) ____
Complexity (1-10) ____ Overall Impression (1-10) ____
SCORE ____

Origin Series Kentucky Straight Bourbon Whiskey

Proof: 96
Age: 6 years
Type: Straight
Style: Traditional
Mash Bill: 60% corn, 36% rye, 4% malted barley
Color: Straw
Price: $$

Mike

Nose: Vanilla, cherries, baking spices, and oak.
Taste: Vanilla, cherries, cinnamon, nutmeg, and oak.
Finish: Medium long with oak and spice.

Susan

Nose: Crème brulee, light citrus and berries, a touch of spice.
Taste: Chocolate covered caramel biscuit with cinnamon and oak.
Finish: Long and dry.

Notes

The Origin Series is the first collection of releases from Bardstown Bourbon Company make exclusively from whiskey distilled at its facility and not blended with other sourced whiskeys.

My Score

Notes: ______________________________

____

____

Appearance (1-5) ____ Nose (1-25) ____
Taste (1-25) ____ Finish (1-25) ____
Complexity (1-10) ____ Overall Impression (1-10) ____

SCORE ____

Phifer Pavitt Reserve

Proof: 107
Age: 5 years
Type: Straight
Style: Finished, Cask Strength
Mash Bill: 84% corn, 8% rye, 8% malted barley
Color: Amber
Price: $$$$

Mike

Nose: Caramel and milk chocolate with caramel and oak.
Taste: Caramel, raspberries and currants, some white pepper and oak.
Finish: Long, dry with oak and pepper.

Susan

Nose: Caramel and dark berries. Fruit influence of the secondary barrel.
Taste: Caramel, marzipan, honey, new leather, rip berries.
Finish: Starts sweet and dries to rich toffee.

Notes

Sourced 9-year-old Tennessee straight bourbon aged an additional 18 months in California cabernet casks. Limited, but high-end whiskey bars may still have some.

My Score

Notes: ______________________________

____

____

Appearance (1-5) ____ Nose (1-25) ____
Taste (1-25) ____ Finish (1-25) ____
Complexity (1-10) ____ Overall Impression (1-10) ____
SCORE ____

Bardstown Bourbon Company Additional Releases

Name: ______________________________

Distillery: ______________________________

Proof: ____________ **Mash Bill:** ____________
Age: ____________ ____________
Type: ____________ **Color:** ____________
Style: ____________ **Price:** ____________

My Score

Notes: ______________________________

____

____

Appearance (1-5) ___ Nose (1-25) ___
Taste (1-25) ___ Finish (1-25) ___
Complexity (1-10) ___ Overall Impression (1-10) ___
SCORE ___

Name: ______________________________

Distillery: ______________________________

Proof: ____________ **Mash Bill:** ____________
Age: ____________ ____________
Type: ____________ **Color:** ____________
Style: ____________ **Price:** ____________

My Score

Notes: ______________________________

____

____

Appearance (1-5) ___ Nose (1-25) ___
Taste (1-25) ___ Finish (1-25) ___
Complexity (1-10) ___ Overall Impression (1-10) ___
SCORE ___

Beam Suntory

www.beamdistilling.com, www.makersmark.com

Bourbon Distilleries:

- James B. Beam, Clermont, Ky.
- Jim Beam, Boston, Ky.
- Maker's Mark, Loretto, Ky.

The Beam family has been distilling in Kentucky since 1795. James B. Beam opened a distillery at the end of prohibition and Jim Beam Bourbon was introduced at that time. T. Jeremiah Beam, the son of James B. Beam, started as treasurer of James Beam Distilling Company in 1934 and it opened the distillery in Clermont, Kentucky. In 1944, he became the president of the company. James B. Beam passed away in 1947.

In 1954, to expand their production, the company acquired the Churchill distillery in Boston, Kentucky. In 1967, American Tobacco Company purchased the company. T. Jeremiah Beam passed away in 1977.

In 1987, James Beam Distilling Company acquired National Distilleries and the Old Grand Dad and Old Crow Brands. In 1992, the company introduced the "small batch" collection of Bourbons based upon Booker's Bourbon, an unfiltered, barrel proof Bourbon that Booker Noe had been giving as gifts since 1989. Beam was purchased in 2014 by Suntory of Japan.

While Beam Suntory does not release its mash bills, the website Modern Thirst (https://modernthirst.com) has been tracking down grain recipes for several years. According to the site, but unconfirmed by the company, the mash bill for most of the Jim Beam products is 75% corn, 13% rye, 12% malted barley. The mash bill for the "high rye" bourbons such as Basil Hayden and Old Grand-Dad is 63% corn, 27% rye, 10% malted barley. The Maker's Mark mash bill is 70% corn, 16% wheat, 14% malted barley. We offer these as guidelines only.

Baker's

Proof: 107
Age: 7 years
Type: Straight
Style: Small Batch
Mash Bill: Not available
Color: Straw
Price: $$$

Mike

Nose: A hint of chocolate and caramel with ripe apple fruit.
Taste: Apples and caramel with some peppery spices. Water brings out the caramel and chocolate.
Finish: Very quickly gets astringent with an alcohol burn. Then gets a little sweet at the very end. Water helps the finish.

Susan

Nose: Vanilla, dash of cocoa, ripe banana.
Taste: Cocoa evolves into saddle leather and caramel apples. Creamy mouthfeel.
Finish: Sweetly spicy and smooth. No burn.

Notes

Part of the Jim Beam Small Batch Collection. Excellent with chocolate desserts.

My Score

Notes: ____

____

____

Appearance (1-5) ____ Nose (1-25) ____
Taste (1-25) ____ Finish (1-25) ____
Complexity (1-10) ____ Overall Impression (1-10) ____

SCORE ____

Basil Hayden's

Proof: 80
Age: NAS
Type: Straight
Style: Small Batch
Mash Bill: High rye
Color: Light Straw
Price: $$

Mike

Nose: Very light nose. A little vanilla and caramel with faint oak in the background. Very faint!
Taste: Corn and vanilla, with a bit of spice. No burn and very little heat.
Finish: Very simple and short. The oak lasts longest, giving it a bit of spice.

Susan

Nose: Very light nose, with a little vanilla and corn and light – very light – spice.
Taste: Vanilla and light spices, which matches the nose.
Finish: Smooth finish starts sweet and dries to the faintly spiced oak. Short, though.

Notes

Part of the Jim Beam Small Batch Collection. Why pay a premium for 80 proof bourbon? That's very expensive, though tasty, water!

My Score

Notes: ______________________________

____

____

Appearance (1-5)	____	Nose (1-25)	____
Taste (1-25)	____	Finish (1-25)	____
Complexity (1-10)	____	Overall Impression (1-10)	____

SCORE ____

Booker's

Proof: 126.5
Age: 7 years, 4 months, 14 days
Type: Straight
Style: Cask Strength, Non-Chill Filtered
Mash Bill: Not available
Color: Dark Straw
Price: $$$

Mike

Nose: Corn, vanilla, leather, hint of fruit, a lot of oak.
Taste: Corn, vanilla, leather, ripe apple, pepper, oak.
Finish: Long and dry with oak and pepper.

Susan

Nose: Toffee, raisins, dark chocolate, nutmeg.
Taste: Matches the nose. Add water and the fruit sweetens to apples and berries.
Finish: Long and complex. Dries to leather and oak.

Notes

Kentucky Tea Batch, 2022-3. Beam typically releases four batches of Booker's per year. The name refers to Booker Noe pouring enough of his bourbon into water that it looked like tea and he drank it with meals.

My Score

Notes: ______________________________

____

____

Appearance (1-5) ____ Nose (1-25) ____
Taste (1-25) ____ Finish (1-25) ____
Complexity (1-10) ____ Overall Impression (1-10) ____

SCORE ____

Devil's Cut

Proof: 90
Age: NAS
Type: Straight
Style: Special Finish
Mash Bill: Not available
Color: Straw
Price: $

Mike

Nose: Vanilla and a bit of pitted fruit – cherry or date. A little leather and tobacco with oak.
Taste: Oak. Lots of oak making it almost too bitter. Caramel and fruit hide in the background.
Finish: Long and woody. Almost astringent.

Susan

Nose: Vanilla, black pepper, ripe banana and touch of caramel and leather.
Taste: Cinnamon spice with vanilla notes. Reminds me of red hot imperial candies.
Finish: Sweet cinnamon lingers in a fairly long finish showing pepper at the end.

Notes

The name refers to the bourbon that soaked into and is rinsed from the red layer of the barrel and added to the dumped whiskey. Opposite of the "angels' share" that evaporates during aging.

My Score

Notes: ____________________

____

____

Appearance (1-5) ____ Nose (1-25) ____
Taste (1-25) ____ Finish (1-25) ____
Complexity (1-10) ____ Overall Impression (1-10) ____

SCORE ____

Hardin's Creek – Colonel James B. Beam

Proof:	108
Age:	2 years
Type:	Straight
Style:	Traditional
Mash Bill:	Not available
Color:	Light Straw
Price:	$$$

Mike

Nose: Corn – lots of it – vanilla, hint of oak.
Taste: Corn, vanilla, pepper, hint of oak.
Finish: Short with corn and vanilla.

Susan

Nose: Vanilla, fruit cocktail, white chocolate, pinch of nutmeg.
Taste: Vanilla cupcake with vanilla icing, dried cherries, light baking spice.
Finish: Spicy.

Notes

Water doesn't change it. Rather pricey for 2 years old, but also pretty complex for its age.

My Score

Notes: ____________________________________

____

____

Appearance (1-5) ___ Nose (1-25) ___
Taste (1-25) ___ Finish (1-25) ___
Complexity (1-10) ___ Overall Impression (1-10) ___

SCORE ___

Hardin's Creek – Jacob's Well

Proof:	108
Age:	15 years, 4 months
Type:	Straight
Style:	Extra Aged
Mash Bill:	Not available
Color:	Amber
Price:	$$$$

Mike

Nose: Caramel, apples, leather, white pepper, and oak.

Taste: Caramel, apples, white pepper, and oak.

Finish: Long with oak, spice, and lingering apple.

Susan

Nose: Vanilla and sweet nuts with some faint ripe apple and a suggestion of cinnamon.

Taste: Oak and leather with some apple and light pepper.

Finish: Long and balanced.

Notes

Water very effectively amplifies the apple note.

My Score

Notes: ____

____

____

Appearance (1-5) ____ Nose (1-25) ____

Taste (1-25) ____ Finish (1-25) ____

Complexity (1-10) ____ Overall Impression (1-10) ____

SCORE ____

Jim Beam

Proof: 80
Age: NAS
Type: Straight
Style: Traditional
Mash Bill: Not available
Color: Light Straw
Price: $

Mike

Nose: Corn and vanilla. Very light nose. A hint of honey.
Taste: Very little heat. A bit thin and watery. Corn, vanilla, and oak. Very little spice, but what is there is sweet.
Finish: Not complex. A little sweet vanilla and oak tannins.

Susan

Nose: Caramel corn and pear with a dash of cinnamon. Not complex.
Taste: Sweet vanilla and frosted corn flakes. Not complex, but a pleasant sip.
Finish: Short, with a distinct cinnamon note at the very end.

Notes

Largest selling bourbon in the world. While there is no age statement on the label, the Jim Beam Black at 8 years states that it is "double aged." So is White Label 4 years old?

My Score

Notes: ____

____

____

Appearance (1-5)	____	Nose (1-25)	____
Taste (1-25)	____	Finish (1-25)	____
Complexity (1-10)	____	Overall Impression (1-10)	____

SCORE ____

Jim Beam Black

Proof: 86
Age: 8 years
Type: Straight
Style: Traditional
Mash Bill: Not available
Color: Golden Straw/Pale Gold
Price: $

Mike

Nose: Caramel corn and oak with a whisper spice – pepper.
Taste: Nice mouthfeel with a little heat and no burn. Caramelized pears and honey with some sweet spices – nutmeg and allspice.
Finish: long and sweet with spices and oak.

Susan

Nose: Vanilla/caramel with a dash of pepper and cinnamon.
Taste: Caramel, apple, and licorice. Cinnamon fades to black pepper.
Finish: Long with fruit at the beginning evolving into dry pepper spice at the end.

Notes

Other than for sipping on its own, try it with a cigar or pipe. Excellent value.

My Score

Notes: ______________________________

____

____

Appearance (1-5) ____ Nose (1-25) ____
Taste (1-25) ____ Finish (1-25) ____
Complexity (1-10) ____ Overall Impression (1-10) ____
SCORE ____

Jim Beam Bonded

Proof: 100
Age: NAS, at least 4 years
Type: Straight
Style: Bonded
Mash Bill: Not available
Color: Light Amber
Price: $$

Mike

Nose: Caramel and vanilla with some corn and oak. Very simple, but nice, nose.
Taste: Candy corn and pepper with some fine tobacco and oak.
Finish: Starts sweet with caramel/vanilla and lasts a long time drying to oak tannins.

Susan

Nose: Bananas, whiff of cinnamon. Vanilla.
Taste: Sweet vanilla and a little cherry. Sweet spices include nutmeg.
Finish: Long, sweet, ending in sweet oak.

Notes

Nice addition to the small, but welcome, band of bottled-in-bond bourbons. Good value.

My Score

Notes: ______________________________

____

____

Appearance (1-5)	____	Nose (1-25)	____
Taste (1-25)	____	Finish (1-25)	____
Complexity (1-10)	____	Overall Impression (1-10)	____

SCORE ____

Jim Beam Distiller's Masterpiece

Proof: 100
Age: NAS
Type: Not Straight
Style: Special Finish
Mash Bill: Not available
Color: Dark Straw
Price: $$$$

Mike

Nose: Chocolate and caramel dominate the nose.
Taste: Fruity with berries and corn, some caramel and oak.
Finish: Long. Starts sweet but dries out to oak and pepper finish.

Susan

Nose: Caramel and stir-fry vegetables.
Taste: Much sweeter than the nose! Caramel, orchard fruit, baking spices and some toasted marshmallow.
Finish: Warm, smooth, long, lingering.

Notes

Beam's entry into the ultra premium bourbon market. Retailing around $200 per bottle, the whiskey is finished in PX sherry casks. Advertising says it is "extra aged" but there is no age statement.

My Score

Notes: ____________________________________

____

____

Appearance (1-5) ____ Nose (1-25) ____
Taste (1-25) ____ Finish (1-25) ____
Complexity (1-10) ____ Overall Impression (1-10) ____

SCORE ____

Jim Beam Repeal Batch

Proof:	86
Age:	NAS
Type:	Straight
Style:	Non-Chill Filtered
Mash Bill:	Not available
Color:	Pale Straw
Price:	$

Mike

Nose: Vanilla and fruit – apples, pears – with a little oak.

Taste: Vanilla cream and apples with a little baking spice and oak.

Finish: Medium long with oak and spice.

Susan

Nose: Ripe apples and vanilla with a whiff of oak.

Taste: Vanilla, corn, more apple, and some floral notes.

Finish: Dries nicely to some oak with a dash of pepper at the end.

Notes

Not complex, but a sippable bargain.

My Score

Notes: ______________________________

____

____

Appearance (1-5) ____ Nose (1-25) ____

Taste (1-25) ____ Finish (1-25) ____

Complexity (1-10) ____ Overall Impression (1-10) ____

SCORE ____

Jim Beam Signature Craft

Proof:	86
Age:	12 years
Type:	Straight Bourbon
Style:	Small Batch, Extra Aged
Mash Bill:	Not available
Color:	Straw
Price:	$$

Mike

Nose: Caramel corn and apricots.

Taste: Caramel and fruit – apricots and dates, with a bit of spice.

Finish: Long. Starts sweet but gets some nice oak and spice as it lasts.

Susan

Nose: Vanilla and graham crackers, some ripe apple.

Taste: Oaky entry, then vanilla and a little leather and tobacco.

Finish: Dry, but with a sweet caramel note at the end.

Notes

Very pleasant sip. Look for this in well-stocked bourbon bars. Beam is releasing several expressions in the Signature Craft series.

My Score

Notes: ____

____

____

Appearance (1-5)	____	Nose (1-25)	____
Taste (1-25)	____	Finish (1-25)	____
Complexity (1-10)	____	Overall Impression (1-10)	____

SCORE ____

Jim Beam 12-Year-Old Signature Craft

Proof: 86
Age: 12 years
Type: Straight
Style: Small Batch, Extra Aged
Mash Bill: Not available
Color: Straw
Price: $$$

Mike

Nose: Very light on the nose. Caramel and a hint of fruit – pears.
Taste: Sweet caramel and fruit – apples and pears – hint of oak.
Finish: Short and sweet. not very interesting.

Susan

Nose: Faint vanilla. Not much else. Perhaps a bit of corn.
Taste: Smooth for its age. Sweet corn and light vanilla on the palate.
Finish: Very short. Not hot, though.

Notes

Perhaps a bit pricey for what's in the bottle.

My Score

Notes: ______________________________________

____

____

Appearance (1-5) ____ Nose (1-25) ____
Taste (1-25) ____ Finish (1-25) ____
Complexity (1-10) ____ Overall Impression (1-10) ____

SCORE ____

Jim Beam Single Barrel

Proof:	95
Age:	NAS, bottled 6-18-14
Type:	Straight
Style:	Single Barrel
Mash Bill:	Not available
Color:	Light Straw
Price:	$$

Mike

Nose: Corn, vanilla and floral notes with a hint of oak.

Taste: Sweet corn and vanilla, pepper spice and fine tobacco.

Finish: Starts peppery, but dries out with oak tannins for a very nice finish.

Susan

Nose: Very floral, lots of vanilla.

Taste: Nutty. Snickers Bar in a glass.

Finish: Warm with no burn. Long and spicy.

Notes

Fascinatingly different from the regular Jim Beam profile. Goes well with a pipe of tobacco.

My Score

Notes: ______________________________

____

____

Appearance (1-5) ____ Nose (1-25) ____

Taste (1-25) ____ Finish (1-25) ____

Complexity (1-10) ____ Overall Impression (1-10) ____

SCORE ____

Knob Creek

Proof:	100
Age:	9 years
Type:	Straight
Style:	Small Batch
Mash Bill:	Not available
Color:	Light Amber
Price:	$$

Mike

Nose: Caramel corn and honey with oak and leather. Just a hint of fruit – apricot.

Taste: A bit hot with caramel and pepper spice.

Finish: After the heat, it gets some nice fruit and caramel and lasts a fair amount of time, Getting drier with oak tannins.

Susan

Nose: Vanilla, light spice and a little sweet cheery fruit.

Taste: Vanilla predominates with caramel showing up mid-palate. Pepper and a flash of fruit.

Finish: Medium long. Fruit quickly fades to dry, almost hot, finish.

Notes

Part of the Jim Beam Small Batch Collection. Look for retailers' single barrel selections. They are often superior to the standard bottling.

My Score

Notes: ______________________________________

____

____

Appearance (1-5) ____ Nose (1-25) ____

Taste (1-25) ____ Finish (1-25) ____

Complexity (1-10) ____ Overall Impression (1-10) ____

SCORE ____

Maker's Mark

Proof: 90
Age: NAS
Type: Straight
Style: Wheated
Mash Bill: 70% corn, 16% red winter wheat, 14% malted barley
Color: Dark Straw
Price: $$

Mike

Nose: Vanilla and caramel with a bit of apple fruit and oak.
Taste: A bit thin, but sweet with caramel and green apples. Pecans and oak wood.
Finish: Very nice oak tannins and apple. Lasts longer than the thin taste would indicate.

Susan

Nose: Light caramel and vanilla with faint pear note and oak.
Taste: Sweet corn and nuts, light vanilla and oak. Pleasant mouthfeel, but a bit of a burn.
Finish: Medium long and peppery with quite a bit of oak tannin at the end.

Notes

A good, solid bourbon to sip neat, but also works well in a variety of cocktails.

My Score

Notes: ____________________

____

____

Appearance (1-5) ___ Nose (1-25) ___
Taste (1-25) ___ Finish (1-25) ___
Complexity (1-10) ___ Overall Impression (1-10) ___

SCORE ___

Maker's Mark Cask Strength

Proof:	113.2
Age:	NAS, but Bill Samuels told us 6 years
Type:	Straight
Style:	Barrel Proof, Wheated
Mash Bill:	70% corn, 16% red winter wheat, 14% malted barley
Color:	Amber
Price:	$

Mike

Nose: Vanilla, caramel, apples and pecans. Not complex, but nice.

Taste: Very nice chewy mouthfeel with warmth in the throat. Caramel and apples with a hint of honey and oak.

Finish: Long and nice with oak and a hint of pepper.

Susan

Nose: Lots of oak, with ripe apples and notes of caramel.

Taste: Sweet and smoky, like good barbecued pork, or even bacon!

Finish: Sweet and oaky, but no woody bitterness.

Notes

A welcome addition to the Maker's portfolio. Available in 375ml and 750ml bottles.

My Score

Notes: ______________________________

____

____

Appearance (1-5)	____	Nose (1-25)	____
Taste (1-25)	____	Finish (1-25)	____
Complexity (1-10)	____	Overall Impression (1-10)	____

SCORE ____

Maker's 46

Proof: 94
Age: NAS
Type: Straight
Style: Wheated, Special Finish
Mash Bill: 70% corn, 16% red winter wheat, 14% malted barley
Color: Amber
Price: $$

Mike

Nose: Vanilla and caramel with oak wood. A hint of fruit – apples and cherry.
Taste: Fruity – apples and cherries with vanilla and oak. A bit of peppery spice.
Finish: Dry oak and spice, but rather short. Ends on the bitter side.

Susan

Nose: Herbal, grassy with prominent wood and dried fruit and nuts.
Taste: Nutty with a dash of cinnamon and dried fruit. But oak predominates.
Finish: Short and warm. Lots of black pepper and and woody tannins.

Notes

Note this expression of Maker's is higher proof than the regular. After the regular warehousing, it is aged several months longer in barrels containing seared French oak staves.

My Score

Notes: ______________________________
____
____

Appearance (1-5) ____ Nose (1-25) ____
Taste (1-25) ____ Finish (1-25) ____
Complexity (1-10) ____ Overall Impression (1-10) ____

SCORE ____

Maker's Mark 101 Traveler's Exclusive

Proof: 101
Age: NAS
Type: Straight
Style: Wheated
Mash Bill: 70% corn, 16% red winter wheat, 14% malted barley
Color: Straw
Price: $$

Mike

Nose: Vanilla, apples, hazelnuts, and oak.
Taste: Vanilla, hazelnuts, and hint of apple, pepper.
Finish: Long with oak and pepper.

Susan

Nose: Vanilla, baking spices, baked apples.
Taste: Caramel apple, vanilla cream, cinnamon, nutmeg, sweet nuts.
Finish: Smooth and sweet with a touch of spice.

Notes

Sold in 1 liter bottles at Global Travel Retail shops. Not available in the U.S., so look for it when traveling internationally.

My Score

Notes: ______________________________

____

____

Appearance (1-5) ____ Nose (1-25) ____
Taste (1-25) ____ Finish (1-25) ____
Complexity (1-10) ____ Overall Impression (1-10) ____
SCORE ____

Maker's Mark Heart Release 2024

Proof: 111.7
Age: NAS
Type: Straight
Style: Wheated, Finished
Mash Bill: 70% corn, 16% red winter wheat, 14% malted barley
Color: Dark Straw
Price: $$$

Mike

Nose: Vanilla, caramel, maple, oak.
Taste: Vanilla, caramel, maple, milk chocolate, pepper, oak.
Finish: Long with oak, pepper, and lingering chocolate.

Susan

Nose: Chocolate, light oak, vanilla.
Taste: Chocolate, caramel, cinnamon sugar, apple, graham cracker.
Finish: Long and balanced. Chocolate gets the last word.

Notes

Part of the annual Wood Finishing Series. Water amplifies the chocolate.

My Score

Notes: ______________________________

____

____

Appearance (1-5) ____ Nose (1-25) ____
Taste (1-25) ____ Finish (1-25) ____
Complexity (1-10) ____ Overall Impression (1-10) ____

SCORE ____

Old Crow

Proof: 80
Age: 3 years
Type: Straight
Style: Traditional
Mash Bill: Not available
Color: Pale Yellow
Price: $

Mike

Nose: Very light nose. Vanilla and corn and not much else.
Taste: Very thin, but some nice vanilla and corn flavor. Not hot at all.
Finish: Longer than expected with sweet vanilla at first, turning into a nice oak dryness.

Susan

Nose: Simple. Sweet corn and hay. Whiff of green apple.
Taste: Corn, nuts, malted barley. No depth.
Finish: Starts sweet, but quickly fades to a slightly peppery, dry finish.

Notes

Lawn mower bourbon with ice and soda? Alas, this historic brand is a pale version of the original.

My Score

Notes: ______________________________

____

____

Appearance (1-5) ____ Nose (1-25) ____
Taste (1-25) ____ Finish (1-25) ____
Complexity (1-10) ____ Overall Impression (1-10) ____
SCORE ____

Old Crow Reserve

Proof:	86
Age:	4 years
Type:	Straight
Style:	Traditional
Mash Bill:	Not available
Color:	Dark Straw
Price:	$

Mike

Nose: Corn and light caramel with some leather and oak.

Taste: A little thin, but some nice caramel and corn with a bit of pepper spice and tobacco.

Finish: Very nice with dry oak and pepper spice.

Susan

Nose: Straightforward. Caramel, light vanilla and oak.

Taste: Vanilla again, with a faint note of cocoa and sweet spice.

Finish: Short and smooth with a touch of pepper.

Notes

Very nice as an inexpensive bourbon to use with mixers.

My Score

Notes: ____________________________________

____

____

Appearance (1-5) ____ Nose (1-25) ____

Taste (1-25) ____ Finish (1-25) ____

Complexity (1-10) ____ Overall Impression (1-10) ____

SCORE ____

Old Grand-Dad

Proof: 80
Age: NAS
Type: Straight
Style: Traditional
Mash Bill: High rye
Color: Straw
Price: $

Mike

Nose: Fruit and caramel – cherries and citrus. Just a bit of oak in the background.
Taste: Cherries and dates. Quite a bit more complex than expected. Caramel, oak and spice – cinnamon.
Finish: A bit short, but pleasant with oak and and spice.

Susan

Nose: Sweet vanilla, almost crème brûlée with buttery pear.
Taste: Surprisingly herbal after the nose. Vanilla and some cinnamon and caramel.
Finish: Medium length and warm. No burn.

Notes

A great bourbon to enjoy with a meal. President Harry Truman had a shot with breakfast every morning. Excellent bargain.

My Score

Notes: ________________________________

____

____

Appearance (1-5)	____	Nose (1-25)	____
Taste (1-25)	____	Finish (1-25)	____
Complexity (1-10)	____	Overall Impression (1-10)	____

SCORE ____

Old Grand-Dad Bonded

Proof: 100
Age: NAS, but at least 4 years
Type: Straight
Style: Bonded
Mash Bill: High rye
Color: Pale Amber
Price: $

Mike

Nose: Cherries and orange peel with some nutmeg and cinnamon.
Taste: Nice mouthfeel with no burn and a little heat. Caramel and citrus fruit with spices – cinnamon.
Finish: Long and spicy with sweet caramel and dry oak.

Susan

Nose: Chocolate and caramel with honey and citrus and a little sweet spice.
Taste: Toffee and candied nuts. Rich vanilla and caramel mid-palate with orange and cinnamon notes. Rich mouthfeel.
Finish: Smooth and sweet. Semi-dry ending.

Notes

A very good sip with food or a cigar. Lovely with chocolate, too. Excellent value.

My Score

Notes: ______________________________

____

____

Appearance (1-5) ____ Nose (1-25) ____
Taste (1-25) ____ Finish (1-25) ____
Complexity (1-10) ____ Overall Impression (1-10) ____

SCORE ____

Old Grand-Dad 114

Proof: 114
Age: NAS
Type: Straight
Style: High proof
Mash Bill: High rye
Color: Deep Amber
Price: $$

Mike

Nose: Very complex. Cherries, dates and a bit orange peel, caramel and sweet spice. I could spend an hour just smelling this bourbon.

Taste: Almost chewy. Nice thick mouthfeel with caramel and fruit- citrus and apricots with some leather and tobacco.

Finish: Nice and long, starts sweet with fruit but gets dry with oak tannins.

Susan

Nose: Vanilla, dash of cocoa, fruit basket and sweet baking spices. Fair amount of alcohol, which is not surprising given the proof, but not unpleasant by any means.

Taste: Baked crème caramel. Touch of water brings out the vanilla and orange peel while taming the tannins.

Finish: Citrus fruit dries into light tannins.

Notes

Water weakens the flavors without opening new ones.

My Score

Notes: ______________________________

____

____

Appearance (1-5) ____ Nose (1-25) ____
Taste (1-25) ____ Finish (1-25) ____
Complexity (1-10) ____ Overall Impression (1-10) ____

SCORE ____

Old Tub

Proof: 100
Age: 4 years
Type: Straight Bourbon
Style: Bottled-in-Bond
Mash Bill: Not available
Color: Dark Straw
Price: $

Mike

Nose: Corn, vanilla and a bit of tobacco and leather.
Taste: Corn and caramel with some pepper spice.
Finish: Nice – some pepper, spice, and oak.

Susan

Nose: Vanilla, citrus, buttered popcorn.
Taste: Fruit cake with caramel and a touch of mint. Very smooth.
Finish: Medium long and warm with a dash black pepper/ sweet oak at the end.

Notes

This can only be purchased at the Jim Beam American Stillhouse located on the distillery grounds in Clermont, Kentucky. Worth getting a 375ml bottle (its only packaging) when you are visiting.

My Score

Notes: ______________________________

____

____

Appearance (1-5) ____ Nose (1-25) ____
Taste (1-25) ____ Finish (1-25) ____
Complexity (1-10) ____ Overall Impression (1-10) ____
SCORE ____

Beam Suntory Additional Releases

Name: ______________________________

Distillery: ______________________________

Proof:	__________	**Mash Bill:**	__________
Age:	__________		__________
Type:	__________	**Color:**	__________
Style:	__________	**Price:**	__________

My Score

Notes: ______________________________

____

____

Appearance (1-5)	___	Nose (1-25)	___
Taste (1-25)	___	Finish (1-25)	___
Complexity (1-10)	___	Overall Impression (1-10)	___

SCORE ___

Name: ______________________________

Distillery: ______________________________

Proof:	__________	**Mash Bill:**	__________
Age:	__________		__________
Type:	__________	**Color:**	__________
Style:	__________	**Price:**	__________

My Score

Notes: ______________________________

____

____

Appearance (1-5)	___	Nose (1-25)	___
Taste (1-25)	___	Finish (1-25)	___
Complexity (1-10)	___	Overall Impression (1-10)	___

SCORE ___

Brown-Forman

www.oldforester.com, www.woodfordreserve.com

Bourbon Distilleries:

- Brown-Forman Distillery, Shively, Ky.
- Old Forester Distillery, Louisville, Ky.
- Woodford Reserve Distillery, Woodford County, Ky.

George Garvin Brown and his half-brother J.T.S. Brown, Jr. entered the whiskey business in 1870 and created the Old Forrester brand as the first Bourbon to be sold only by the bottle. This whiskey was aimed at sales to the medical market and sold bottled to guarantee the quality of the whiskey being prescribed by doctors to their patients. Doctor William Forrester endorsed the product and allowed Brown to use his name on the label. When Dr. Forrester retired, the second "r" was dropped from the name. J.T.S. Brown left the company a few years later to form his own company and George Garvin Brown forms what would eventually become Brown-Forman.

Brown-Forman was one of the six companies with a license to sell "medicinal alcohol" during prohibition. It was during prohibition when the company acquired the Early Times Brand and stocks of whiskey. Early Times is still a Bourbon in the Japanese market, but except for limited releases, it is now "Kentucky Style" whiskey (aged in used cooperage) in the U.S.

In 1997 Brown-Forman opened the Woodford Reserve Distillery and introduced Woodford Reserve Bourbon. The Pepper family founded the distillery in the early 19th century. In 1870s, James E. Pepper as a minor owned the distillery and his guardian was E. H. Taylor, Jr. In 1878 financial difficulties forced the sale of the distillery to Labrot and Graham. Brown-Forman acquired the distillery and its stocks in 1940 but closed and sold it to a local farmer in 1973. They purchased the distillery back in 1995 and opened it in 1997.

In addition to the bourbons listed here, a special Master's Collection edition of Woodford Reserve is released annually. Some have special finishes, such as having been aged in wine cooperage. Others have special recipes, such as using chocolate rye or four grains.

Coopers' Craft

Proof: 82.2
Age: NAS
Type: Straight
Style: Toasted Wood Whiskey
Mash Bill: 75% corn, 20% rye, 10% malted barley
Color: Light Straw
Price: $

Mike

Nose: Caramel and fruit with some oak and leather.
Taste: Sweet caramel and apple fruit with pepper spice.
Finish: Short, but sweet and fruity. Brandy-like.

Susan

Nose: Fruit salad- bananas, orange. Caramel and sweet spice.
Taste: Drier on palate than nose. Lots of oaky spice, some apple and vanilla.
Finish: Medium and sweet, ending in oak.

Notes

This is Brown-Forman's first new whiskey since the introduction of Woodford Reserve in 1996. The company is the only whiskey distillery in the world that owns its own cooperage, and this new bourbon was aged in barrels toasted differently from those used for B-F's other bourbons.

My Score

Notes: ______________________________

____

____

Appearance (1-5) ____ Nose (1-25) ____
Taste (1-25) ____ Finish (1-25) ____
Complexity (1-10) ____ Overall Impression (1-10) ____

SCORE ____

Coopers' Craft Barrel Reserve

Proof: 100
Age: NAS
Type: Straight
Style: Traditional
Mash Bill: 75% corn, 15% rye, 10% malted barley
Color: Light Amber
Price: $$

Mike

Nose: Caramel and toffee, a bit of pitted fruit and oak.
Taste: Caramel and apricots with some pepper spice and oak.
Finish: Long and peppery.

Susan

Nose: Caramel and orchard fruit with a bit of buttered popcorn.
Taste: Caramel, almonds, ripe fruit (some banana), and a dash of cinnamon.
Finish: Warm and spicy.

Notes

Water reveals apple and pear notes.

My Score

Notes: ____________________________________

____

____

Appearance (1-5) ____ Nose (1-25) ____
Taste (1-25) ____ Finish (1-25) ____
Complexity (1-10) ____ Overall Impression (1-10) ____

SCORE ____

King of Kentucky

Proof: 131
Age: 15 years
Type: Straight
Style: Barrel Proof, Non-Chill Filtered
Mash Bill: 79% corn, 11% rye, 10% malted barley
Color: Dark Amber
Price: $$$$

Mike

Nose: Crème brulee, oak, and a hint of fruit.
Taste: Oak tannins up front with apples, pepper, and caramel. Water reduces the tannins and enhances the fruit and caramel.
Finish: Long and oaky. Water makes the finish less bitter.

Susan

Nose: Crème brulee, sweet cherries, almonds, banana chips.
Taste: Marzipan, milk chocolate, dark fruit, some oak. Water richens the mouthfeel.
Finish: Very smooth for the proof. Dries to spicy oak.

Notes

Barrel 8/27 with 2,100 bottles produced. Annual limited release only available in Kentucky. Age can range from 10 years upward and proof between 125 and 135.

My Score

Notes: ____

____

____

Appearance (1-5) ____ Nose (1-25) ____
Taste (1-25) ____ Finish (1-25) ____
Complexity (1-10) ____ Overall Impression (1-10) ____

SCORE ____

Old Forester

Proof: 86
Age: NAS
Type: Straight Bourbon
Style: Straight
Mash Bill: 72% corn, 18% rye, 10% malted barley
Color: Dark Straw
Price: $

Mike

Nose: Dark fruit – apricots and ripe apples with caramel, fine leather and pepper spice. Just a hint of oak.
Taste: Very nice mouthfeel with no alcohol burn. Apples, caramel, pepper spice, vanilla and hint of oak.
Finish: Long and dry with oak and pepper spice that leads to sweet vanilla.

Susan

Nose: Complex nose with vanilla, caramel, apple with faint nuttiness and oak.
Taste: Caramel and vanilla, with notes of chocolate, leather, apple and hint of tobacco.
Finish: Smooth, long, peppery (but not hot) finish. Sweet vanilla note at the very end.

Notes

A classic. If you are introducing someone to bourbon, this would be a fine one to use as a benchmark. As fine in a cocktail (makes a splendid Old Fashioned) as it is sipped neat or with a single ice cube.

My Score

Notes: ____

____

____

Appearance (1-5) ____ Nose (1-25) ____
Taste (1-25) ____ Finish (1-25) ____
Complexity (1-10) ____ Overall Impression (1-10) ____
SCORE ____

Old Forester 1870

Proof: 90
Age: NAS
Type: Straight Bourbon
Style: Small Batch
Mash Bill: 72% corn, 18% rye, 10% malted barley
Color: Light Amber
Price: $$

Mike

Nose: Very nice. caramel and vanilla with some fruit – berries and apple, sweet spices.
Taste: Nice mouthfeel, almost chewy with sweet caramel and fruit – apple and pear.
Finish: Long and spicy with pepper and oak.

Susan

Nose: Lots of OF's signature caramel plus cherries, pears, sprinkling of nutmeg.
Taste: Rich mouthfeel with smoked toffee and baked fruit – cherry pie.
Finish: Long and warm. No burn. Pleasant spicy ending.

Notes

Attempt to recreate George Garvin Brown's original batch from 1870, the founding date of the distillery, by sourcing three bourbons with different production dates, different barrel entry proofs, and different maturation periods.

My Score

Notes: ______________________________

____

____

Appearance (1-5) ____ Nose (1-25) ____
Taste (1-25) ____ Finish (1-25) ____
Complexity (1-10) ____ Overall Impression (1-10) ____

SCORE ____

Old Forester 1897

Proof: 100
Age: At least 4 years
Type: Straight
Style: Bottled-in-Bond
Mash Bill: 72% corn, 18% rye, 10% malted barley
Color: Light Amber
Price: $$$

Mike

Nose: Bananas, caramel, and vanilla.
Taste: Caramel, bananas, baking spice, Bananas Foster.
Finish: Dry with oak and pepper.

Susan

Nose: Caramel, cinnamon, pears.
Taste: Cinnamon coffee cake with a layer of vanilla icing.
Finish: Dry with oak and pepper. Long and peppery.

Notes

The second release in Brown-Forman's Whiskey Row Series. This one pays homage to the Bottled-in-Bond Act of 1897, America's first consumer protection law, meant to guarantee whiskey quality.

My Score

Notes: ____________________________________

____

____

Appearance (1-5) ____ Nose (1-25) ____
Taste (1-25) ____ Finish (1-25) ____
Complexity (1-10) ____ Overall Impression (1-10) ____

SCORE ____

Old Forester 1910 Old Fine Whisky

Proof:	93
Age:	NAS
Type:	Straight
Style:	Double Barreled
Mash Bill:	72% corn, 18% rye, 10% malted barley
Color:	Amber
Price:	$$$

Mike

Nose: Very brandy-like. Lots of prunes and dates with caramel and oak.

Taste: Very fruity with cherries, dates, caramel, pepper, and oak.

Finish: Starts sweet but dries to oak and pepper.

Susan

Nose: Burnt caramel, toffee, bananas, and smoke.

Taste: Light smoke, caramel, milk chocolate, leather, and a little more smoke.

Finish: Long and peppery.

Notes

After initial aging, the bourbon is put in a second barrel that is charred to the point right before ash forms. Susan's favorite Whiskey Row Series expression.

My Score

Notes: ____

____

____

Appearance (1-5)	____	Nose (1-25)	____
Taste (1-25)	____	Finish (1-25)	____
Complexity (1-10)	____	Overall Impression (1-10)	____

SCORE ____

Old Forester 1920

Proof:	115
Age:	NAS
Type:	Straight
Style:	Cask Strength
Mash Bill:	72% corn, 18% rye, 10% malted barley
Color:	Rich Amber
Price:	$$$

Mike

Nose: Apricots and berries, caramel, tobacco and a hint of baking spice.

Taste: Fruit and spice with caramel toffee. Water brings out the fruit and spice.

Finish: Dry oak and spice. Water makes it spicier.

Susan

Nose: Lots of caramel, sweet spices, ripe apple.

Taste: Rich toffee without water. With water, more nuts, apples, spice.

Finish: Long, oaky, spicy.

Notes

Excellent sip, but do add water/ice. The third release in Brown-Forman's Whiskey Row Series. "Prohibition Style" since Brown-Forman was one of six distillery's given a government permit to make "medicinal whiskey." Barrel proof before bottling was typically 120.

My Score

Notes: ____________________

____

____

Appearance (1-5)	___	Nose (1-25)	___
Taste (1-25)	___	Finish (1-25)	___
Complexity (1-10)	___	Overall Impression (1-10)	___

SCORE ___

Old Forester Birthday Bourbon 2013

Proof: 94
Age: 12 years
Type: Straight Bourbon
Style: Small Batch, Extra Aged
Mash Bill: 72% corn, 18% rye, 10% malted barley
Color: Deep Amber
Price: $$$$

Mike

Nose: Rich caramel corn and vanilla. Some fruit – berries and apples, with sweet spices. A bit of fine leather in the background.
Taste: Sweet fruit – apples, berries with caramel and a hint of anise/nutmeg. Some oak wood and leather.
Finish: Long and sweet to start, but dries out nicely with oak tannins and spice.

Susan

Nose: Ripe cherries, vanilla, and sweet spices. Intriguingly complex.
Taste: Rich caramel, vanilla. The cherry in the nose carries over to the palate, joined by cinnamon and buttery corn.
Finish: Long and fruit-filled. Sweet oak tannins and a bit of spicy warmth at the end.

Notes

Brown-Forman releases this mingling of 12-year-old Old Foresters every year in September to commemorate founder George Garvin Brown's birthday (2 September, 1846). They vary from year to year, but are always worth seeking out.

My Score

Notes: ____________________________________

____

____

Appearance (1-5) ____ Nose (1-25) ____
Taste (1-25) ____ Finish (1-25) ____
Complexity (1-10) ____ Overall Impression (1-10) ____

SCORE ____

Old Forester Birthday Bourbon 2019

Proof: 105
Age: 11 years
Type: Straight
Style: Limited Release
Mash Bill: 72% corn, 18% rye, 10% malted barley
Color: Straw
Price: $$$$

Mike

Nose: French vanilla and leather with bananas and oak.
Taste: Vanilla, bananas, leather, oak, white pepper.
Finish: Long and dry with oak and spice.

Susan

Nose: Bananas, cherries, baking spices and lots of caramel and vanilla.
Taste: Fruit and spice, but give it time to open in the glass.
Finish: Long and spicy, with much of the fruit.

Notes

Do add a little water to tame the proof and allow the fruit and caramel to take charge.

My Score

Notes: ____

____

____

Appearance (1-5) ____ Nose (1-25) ____
Taste (1-25) ____ Finish (1-25) ____
Complexity (1-10) ____ Overall Impression (1-10) ____
SCORE ____

Old Forester Signature

Proof:	100
Age:	NAS
Type:	Straight Bourbon
Style:	Straight
Mash Bill:	72% corn, 18% rye, 10% malted barley
Color:	Amber
Price:	$

Mike

Nose: Dark fruit – apricots and dates, apples and caramel. Just a hint of spice and oak. Fine leather and tobacco. Very complicated. The longer it breathes, the more that comes out on the nose.

Taste: Thick and chewy with only a little warmth in the back of the throat. Apples, caramel, cherries and peppery spice with hint of oak.

Finish: Long and pleasing with oak and pepper changing to caramel and vanilla.

Susan

Nose: Complex nose with caramel, vanilla, dark fruit, roasted corn and toasted pecans. Be patient. More layers emerge as it sits in the glass.

Taste: Caramel and vanilla, with notes of bright cherry and apple. Oaky without being woody.

Finish: Vanilla and caramel dominate the finish which shows a peppery, but not hot, note at the end.

Notes

Beautifully balanced. Has the character and complexity of a bourbon twice its price.

My Score

Notes: ______________________________

____

____

Appearance (1-5)	____	Nose (1-25)	____
Taste (1-25)	____	Finish (1-25)	____
Complexity (1-10)	____	Overall Impression (1-10)	____

SCORE ____

Old Forester Single Barrel

Proof: 90
Age: Varies, but over 6 years
Type: Straight Bourbon
Style: Single Barrel
Mash Bill: 72% corn, 18% rye, 10% malted barley
Color: Light Amber
Price: $$$

Mike

Nose: Apricots and peaches with vanilla and caramel. Some sweet spices and oak.
Taste: Sweet fruit – apricots and apples with some cinnamon spice and fine tobacco.
Finish: A little peppery and dry with oak. Lasts a long time.

Susan

Nose: Caramel and candied almonds with ripe cherries. Faint cinnamon notes.
Taste: Caramel and vanilla predominate supported by oak tannins. Fruit and nuts linger in the background.
Finish: Long and warm. Dries to sweet oakiness.

Notes

Private selections for retail stores, restaurants, whiskey clubs, and even individuals. Can have very distinct flavors that may not show up in other Old Forester expressions.

My Score

Notes: ____________________________________
____
____

Appearance (1-5)	____	Nose (1-25)	____
Taste (1-25)	____	Finish (1-25)	____
Complexity (1-10)	____	Overall Impression (1-10)	____

SCORE ____

Woodford Reserve Distiller's Select

Proof: 90.4
Age: NAS
Type: Straight Bourbon
Style: Small Batch
Mash Bill: 72% corn, 18% rye, 10% malted barley
Color: Light Amber
Price: $$

Mike

Nose: Rich caramel. Crème brulee with pitted Fruit – cherries and peaches. Apricots with a hint of oak and leather.
Taste: Caramel and fruit with some sweet spice and oak.
Finish: Long and sweet – caramel and vanilla with only a hint of oak and spice.

Susan

Nose: Intense vanilla and caramel with fruit and flowers, lightly nutty and spicy.
Taste: Creamy caramel and vanilla. The orange emerges as the fruit, with layer of nuttiness and nutmeg.
Finish: Lingering. Warmth ends with sweet vanilla and a whisper of the nutmeg.

Notes

Richly flavored and fine neat or with a splash. Makes an excellent Manhattan paired with the best vermouth you can find.

My Score

Notes: ____

____

____

Appearance (1-5) ____ Nose (1-25) ____
Taste (1-25) ____ Finish (1-25) ____
Complexity (1-10) ____ Overall Impression (1-10) ____

SCORE ____

Woodford Reserve Double Double Oaked

Proof: 90.4
Age: NAS
Type: Straight
Style: Small Batch
Mash Bill: 72% corn, 18% rye, 10% malted barley
Color: Amber
Price: $$$

Mike

Nose: Crème brulee, fruit, oak, and baking spice.
Taste: Toffee and pitted fruit with some sweet spice.
Finish: Starts buttery caramel but dries out with oak and spice.

Susan

Nose: Maple and toffee with baked apples.
Taste: Smoky. Rich caramel, vanilla, dates.
Finish: Long with sweet oak and wood smoke. Spicier and less sweet that the regular Double oak.

Notes

A true after dinner sip. Retailing in Kentucky for about $40 for 375ml bottle, hence the dollar sign number. This is Woodford Double Oak aged for yet another year in heavily-toasted, lightly-charred new oak barrels.

My Score

Notes: ______________________________

____

____

Appearance (1-5) ____ Nose (1-25) ____
Taste (1-25) ____ Finish (1-25) ____
Complexity (1-10) ____ Overall Impression (1-10) ____

SCORE ____

Woodford Reserve Double Oaked

Proof: 90.4
Age: NAS
Type: Straight Bourbon
Style: Small Batch, Extra Aged
Mash Bill: 72% corn, 18% rye, 10% malted barley
Color: Amber
Price: $$$

Mike

Nose: Oak and dark, chewy caramel. Very simple nose.
Taste: Oak wood and bitter tannins. Apples with hint of caramel.
Finish: Long and dry with oak tannins. Only a hint of oak and spice.

Susan

Nose: Oak and dark, rich caramel. Not complex.
Taste: Oaky maple syrup. Some caramel apple.
Finish: Long. Vanilla kicks in at the very end.

Notes

Dumped into a second oak barrel that has been deeply toasted and lightly charred. It is then aged for an additional six months. Add a splash of water to tame the tannins.

My Score

Notes: ____________________

____

____

Appearance (1-5)	___	Nose (1-25)	___
Taste (1-25)	___	Finish (1-25)	___
Complexity (1-10)	___	Overall Impression (1-10)	___

SCORE ___

Woodford Reserve Double Oaked Single Barrel

Proof: 90.4
Age: NAS
Type: Straight Bourbon
Style: Single Barrel, Extra Aged
Mash Bill: 72% corn, 18% rye, 10% malted barley
Color: Amber
Price: $$$

Mike

Nose: Caramel toffee and spice. A bit of oak and fruit – apples and apricots.
Taste: Apple pie with caramel – a bit of pepper spice and oak.
Finish: Long. Starts peppery but dries out with oak and hickory nuts.

Susan

Nose: Nutty caramel, brown sugar, oak, and dark cherries.
Taste: Oaky, yeasty, nutty. Rich, chewy mouthfeel and lots of sweet vanilla.
Finish: Very long. Begins with caramel sweetness and dries to light spice and light oak.

Notes

Single Barrel selections are private bottlings for retailers, restaurants, and such. Look for this in a bourbon bar with large number of bourbons and several private bottlings.

My Score

Notes: ______________________________
____
____

Appearance (1-5) ____ Nose (1-25) ____
Taste (1-25) ____ Finish (1-25) ____
Complexity (1-10) ____ Overall Impression (1-10) ____
SCORE ____

Brown-Forman Additional Releases

Name: ______________________________

Distillery: ______________________________

Proof: __________ **Mash Bill:** __________
Age: __________ __________
Type: __________ **Color:** __________
Style: __________ **Price:** __________

My Score

Notes: ______________________________

____

____

Appearance (1-5) ____ Nose (1-25) ____
Taste (1-25) ____ Finish (1-25) ____
Complexity (1-10) ____ Overall Impression (1-10) ____

SCORE ____

Name: ______________________________

Distillery: ______________________________

Proof: __________ **Mash Bill:** __________
Age: __________ __________
Type: __________ **Color:** __________
Style: __________ **Price:** __________

My Score

Notes: ______________________________

____

____

Appearance (1-5) ____ Nose (1-25) ____
Taste (1-25) ____ Finish (1-25) ____
Complexity (1-10) ____ Overall Impression (1-10) ____

SCORE ____

Diageo

www.bulleit.com, www.stitzelwellerdistillery.com

Bourbon Distilleries:

- Bulleit Distilling Co., Shelbyville, Ky.
- Stitzel-Weller Distillery, Shively, Ky.

London-based Diageo PLC is the largest alcoholic beverage producer in the world. Brands include Johnnie Walker Scotch, Tanqueray gin, Smirnoff vodka, and Guinness stout. Its Bulleit bourbons were contract distilled in Kentucky until its new distillery opened in Shelby County 2019. It also owns the Stitzel-Weller Distillery in Louisville, which it had purchased and closed in 1972. (It was reopened it for tours in 2014.)

While Diageo is coy about its mash bills, Modern Thirst lists the recipes for the Bulleit bourbons as 63% corn, 27% rye and 10% malted barley. Also, according to Modern Thirst, most of the Orphan Barrel series are whopping 86% corn, a mere 6% rye, and 8% malted barley.

Barterhouse

Proof: 90.2
Age: 20 years
Type: Not Straight
Style: Extra Aged
Mash Bill: Not available
Color: Light Amber
Price: $$$$

Mike

Nose: Caramel and oak with some leather and spice.
Taste: Caramel and apricots with sweet spices and leather. A bit thin in the mouth.
Finish: Short and dry with oak and a hint of fruit.

Susan

Nose: Sweet oak, ripe peaches, light vanilla. Rather tight.
Taste: Caramel, vanilla, berries, sweet oak. Some leather.
Finish: Short and warm, ending in oak.

Notes

Interestingly, water does not change the profile at all. One of Diageo's "Orphan Barrel" bourbons bottled in Tullahoma, Tennessee.

My Score

Notes: ____

____

____

Appearance (1-5) ____ Nose (1-25) ____
Taste (1-25) ____ Finish (1-25) ____
Complexity (1-10) ____ Overall Impression (1-10) ____

SCORE ____

Blade and Bow

Proof: 92
Age: NAS
Type: Straight
Style: Solera Aged
Mash Bill: Proprietary
Color: Dark Straw
Price: $$

Mike

Nose: Creamy vanilla and sweet spice, ripe apples and pears. Just a hint of oak.

Taste: Apples and vanilla, with sweet spices-nutmeg and cinnamon. Water brings out sweet fruit and spice.

Finish: Fairly short. Starts sweet and gets spicy with some oak tannins. Water lengthens finish.

Susan

Nose: Dark fruit, light mint, and sweet vanilla.

Taste: Pears, sweet licorice, almonds. Biscotti in a glass.

Finish: Sweet and smooth. Just a touch of oak at the end.

Notes

Certainly snifter-worthy. Tiniest bit of water amplifies fruit and oak. This purports to have some of the last of the stock distilled at the legendary Stitzel-Weller Distillery in Shively, Kentucky.

My Score

Notes: ____________________________________

____

____

Appearance (1-5) ____ Nose (1-25) ____
Taste (1-25) ____ Finish (1-25) ____
Complexity (1-10) ____ Overall Impression (1-10) ____

SCORE ____

Blade and Bow 22-Year-Old

Proof: 92
Age: 22 years
Type: Straight
Style: Limited Release
Mash Bill: Proprietary
Color: Amber
Price: $$$$

Mike

Nose: Lots of fruit. Apples and pears with a hint of berries. Vanilla and caramel.
Taste: Nice mouthfeel – chewy and thick. Fruit and spice balanced by some caramel and chocolate.
Finish: Very simple and short. The oak lasts longest, giving it a bit of spice.

Susan

Nose: Very fruity with peaches, chocolate, and sweet oak.
Taste: Fruit forward with notes of cocoa, honey, and caramel.
Finish: Long and warm. Very smooth and sweet.

Notes

Water dilutes the flavor. Do not add water or ice! Retailing for $180 per bottle, if you can find it. Worth splurging on for a shot in a good whiskey bar, where it is more likely to be found than on a store shelf.

My Score

Notes: ____________________________________

____

____

Appearance (1-5) ____ Nose (1-25) ____
Taste (1-25) ____ Finish (1-25) ____
Complexity (1-10) ____ Overall Impression (1-10) ____
SCORE ____

Bulleit

Proof: 90
Age: NAS
Type: Straight Bourbon
Style: Traditional
Mash Bill: Proprietary, but high rye
Color: Straw
Price: $$

Mike

Nose: Cream soda vanilla, light fruit – ripe pears and apples – touch of oak.

Taste: Rich vanilla and fruit – pears and berries with a little sweet spice and oak.

Finish: Long and spicy. Oak tannins turn to sweet spices – clove, cinnamon, allspice.

Susan

Nose: Candied nuts and light caramel with floral and vanilla notes hovering over the oak.

Taste: Cinnamon spice, bright fruit, and vanilla custard. Hint of mint.

Finish: Long, lingering, pumpkin pie spices and a distinctly oaky note.

Notes

A flavorful bourbon that can be enjoyed neat, but will be excellent in traditional bourbon cocktails. Very versatile and a real bargain at its price point.

My Score

Notes: ______________________________

____

____

Appearance (1-5) ____ Nose (1-25) ____
Taste (1-25) ____ Finish (1-25) ____
Complexity (1-10) ____ Overall Impression (1-10) ____

SCORE ____

Bulleit 10-Year-Old

Proof: 91.2
Age: 10 years
Type: Straight Bourbon
Style: Small Batch, Extra Aged
Mash Bill: Proprietary, but high rye
Color: Amber
Price: $$

Mike

Nose: Floral (roses) and vanilla. Pitted fruit- cherries and apricot. Hint of sweet spice cherries and apricot. Hint of sweet spice in the background.
Taste: Nice chewy mouthfeel with sweet apricot and some sweet cinnamon spice.
Finish: Long and dry. Starts sweet, but dried very quickly by the oak. Very pleasant.

Susan

Nose: Fruity and floral (apple, cherry) with almonds and balanced vanilla and caramel notes.
Taste: Creamy on the palate. More sweet almond and faint spices, supported by oak.
Finish: Long and spicy with oak at the end.

Notes

Very flavorful. A splash of water brings out more of floral character. This is a bourbon over which to linger. Sip slowly.

My Score

Notes: ____

____

____

Appearance (1-5) ____ Nose (1-25) ____
Taste (1-25) ____ Finish (1-25) ____
Complexity (1-10) ____ Overall Impression (1-10) ____

SCORE ____

Bulleit Cask Strength

Proof: 119
Age: NAS
Type: Straight
Style: Barrel Strength
Mash Bill: Proprietary, but high rye
Color: Dark Straw
Price: $$

Mike

Nose: Fruit and spice, caramel and a hint of honeysuckle flower.
Taste: Caramel and berries, baking spices and oak.
Finish: Very nice balance and oak dryness and fruit.

Susan

Nose: Bananas Foster, sweet mint, vanilla.
Taste: Loads of vanilla and ripe fruit – berries. Some oaky spice.
Finish: Smooth and warm, even at barrel proof.

Notes

Oddly, water makes it a bit astringent, rather than opening it up.

My Score

Notes: ____

____

____

Appearance (1-5) ____ Nose (1-25) ____
Taste (1-25) ____ Finish (1-25) ____
Complexity (1-10) ____ Overall Impression (1-10) ____

SCORE ____

Forged Oak

Proof: 90.5
Age: 15 years
Type: Straight
Style: Extra Aged
Mash Bill: Proprietary
Color: Dark Straw
Price: $$$

Mike

Nose: Caramel and oak with a hint of tobacco. Some light apricot and honey.
Taste: Very light. Almost a bit thin. Caramel and apricots with some nutmeg and pepper.
Finish: A bit dry with oak tannins. A little short, too.

Susan

Nose: Caramel, sweet tobacco, a little mint, light apricots.
Taste: Caramel and vanilla with candied nuts. But lightweight for 15 years.
Finish: Lingering and sweet until a sharp note of pepper at the very end.

Notes

This is one of the so-called Orphan Barrel releases by Diageo from stores aged at the historic Stitzel-Weller Distillery in Louisville, Kentucky. Originally distilled at the New Bernheim Distillery in Louisville, which used to be owned by Diageo, but is now Heaven Hill's distillery.

My Score

Notes: ______________________________

____

____

Appearance (1-5) ____ Nose (1-25) ____
Taste (1-25) ____ Finish (1-25) ____
Complexity (1-10) ____ Overall Impression (1-10) ____
SCORE ____

I.W. Harper

Proof: 82
Age: NAS
Type: Straight
Style: Traditional
Mash Bill: Proprietary
Color: Light Amber
Price: $$

Mike

Nose: Caramel and citrus fruit. Some honeysuckle and oak.
Taste: Nice mouthfeel. A bit chewy with caramel and honey, orange peel, and tobacco.
Finish: Long and drying with oak tannins and pepper. Pepper increases over the course of the finish.

Susan

Nose: Caramel, light cinnamon, ripe pears.
Taste: Rich mouthfeel. Butter caramel/toffee. Sweet spices and ripe orchard fruits.
Finish: Begins sweet and dries to a light pepper ending.

Notes

Proof is an interesting choice. A couple of degrees stronger than the export version. Label states it was distilled in Kentucky. No source named.

My Score

Notes: ______________________________________
____
____

Appearance (1-5) ____ Nose (1-25) ____
Taste (1-25) ____ Finish (1-25) ____
Complexity (1-10) ____ Overall Impression (1-10) ____

SCORE ____

I.W. Harper Gold Medal

Proof: 80
Age: NAS
Type: Straight Bourbon
Style: Not available
Mash Bill: Proprietary
Color: Light Amber
Price: $$$ (Export only)

Mike

Nose: Citrus fruit and vanilla, with hint of oak and honey spice (cloves?) hiding in the background.

Taste: Very easy to drink with no alcohol burn. Yet flavorful, with sweet corn, apple and sweet clove and nutmeg.

Finish: Long and sweet with spices and fruit. Just a hint of oak.

Susan

Nose: Honey and nuts, with vanilla and orange blossom over the underlying oak.

Taste: Buttered popcorn. Very rich and smooth with citrus and light spiciness.

Finish: Long and warm, but no burn. Ends in honey and fruit.

Notes

Notably flavorful, complex and satisfying for an 80 proof bourbon! Can be enjoyed neat, but would make a fantastic Old Fashioned. Available only in Asian markets.

My Score

Notes: ______________________________

____

____

Appearance (1-5) ____ Nose (1-25) ____
Taste (1-25) ____ Finish (1-25) ____
Complexity (1-10) ____ Overall Impression (1-10) ____

SCORE ____

I.W. Harper 12-Year-Old

Proof: 86
Age: 12 years
Type: Straight Bourbon
Style: Extra Aged
Mash Bill: Proprietary
Color: Amber Shading to Orange
Price: $$$$ (Export only)

Mike

Nose: Citrus, fruit and vanilla with fine leather and oak, plus honey and nutmeg.
Taste: Very easy to drink with no alcohol burn. Caramel and honey with some sweet peach fruit and cinnamon.
Finish: Long finish starts sweet and ends in honey with a nice oak dryness.

Susan

Nose: Well-balance vanilla, honey, and caramel, with prominent citrus.
Taste: Rich toffee, light fruit and hint of leather, with underlying spice.
Finish: Long and warm and buttery, with a pleasantly dry ending.

Notes

Great bourbon to drink with a meal. Works well with savory food. Regrettably, only available in Asian markets.

My Score

Notes: ______________________________

____

____

Appearance (1-5) ____ Nose (1-25) ____
Taste (1-25) ____ Finish (1-25) ____
Complexity (1-10) ____ Overall Impression (1-10) ____

SCORE ____

I.W. Harper 15-Year-Old

Proof:	86
Age:	15 years
Type:	Straight Bourbon
Style:	Extra Aged
Mash Bill:	Proprietary
Color:	Light Amber
Price:	$$$$

Mike

Nose: Caramel and vanilla citrus and apricots.
Taste: Carmel and apricots with nutmeg and tobacco.
Finish: Long. Starts sweet and gets dry with oak and pepper.

Susan

Nose: Butterscotch, caramel, and oak, and fruity spice.
Taste: Lots of vanilla and caramel, caramel corn. Sprinkling of nutmeg.
Finish: Long and dry, with a peppery ending.

Notes

An elegant, snifter-worthy sip. This was Batch SS-1 FGS.

My Score

Notes: ______________________________

____

____

Appearance (1-5) ____ Nose (1-25) ____
Taste (1-25) ____ Finish (1-25) ____
Complexity (1-10) ____ Overall Impression (1-10) ____

SCORE ____

Rhetoric

Proof: 90
Age: 20 years
Type: Straight Bourbon
Style: Extra Aged
Mash Bill: Proprietary
Color: Amber
Price: $$$$

Mike

Nose: Honeysuckle and caramel with ripe pears and oak.
Taste: Sweet caramel and honey with pears, apples and some spice. Nice mouthfeel with no burn and nice warmth in the throat.
Finish: Starts sweet and dries nicely with a long oak finish.

Susan

Nose: Rich caramel, almonds, ripe apricots, buttered corn.
Taste: Ripe apples and peaches on the tongue. Fruit salad with nuts and vanilla. Very chewy.
Finish: Long and warm and buttery, with a pleasantly dry ending.

Notes

As it breathes it opens up with more oak and citrus. Water makes this sip even sweeter, bringing out honey notes. One of the "Orphan Barrel" releases from Diageo. Company says that it comes from "barrels discovered at the back of he rickhouse" on the Stitzel-Weller property in Louisville, Kentucky. It was bottled at George Dickel in Tullahoma, Tennessee.

My Score

Notes: ____

____

____

Appearance (1-5) ____ Nose (1-25) ____
Taste (1-25) ____ Finish (1-25) ____
Complexity (1-10) ____ Overall Impression (1-10) ____

SCORE ____

Diageo Additional Releases

Name: ______________________________

Distillery: ______________________________

Proof: __________ **Mash Bill:** __________

Age: __________ __________

Type: __________ **Color:** __________

Style: __________ **Price:** __________

My Score

Notes: ______________________________

____

____

Appearance (1-5) ____ Nose (1-25) ____

Taste (1-25) ____ Finish (1-25) ____

Complexity (1-10) ____ Overall Impression (1-10) ____

SCORE ____

Name: ______________________________

Distillery: ______________________________

Proof: __________ **Mash Bill:** __________

Age: __________ __________

Type: __________ **Color:** __________

Style: __________ **Price:** __________

My Score

Notes: ______________________________

____

____

Appearance (1-5) ____ Nose (1-25) ____

Taste (1-25) ____ Finish (1-25) ____

Complexity (1-10) ____ Overall Impression (1-10) ____

SCORE ____

Four Roses

www.fourrosesbourbon.com

Bourbon Distilleries:

- Four Roses Distillery, Lawrenceburg, Ky.

Paul Jones created the Four Roses brand after the Civil War in Atlanta, Georgia using whiskey from the Rose distillery in that city. When Prohibition made Jones leave the state in 1884, he moved to Louisville, Ky., bringing his brand with him. Paul Jones dies in 1895 and the business passes to his nephew. In 1922 The Paul Jones Company purchases the Frankfort Distilleries Company and their license to sell medicinal whiskey during Prohibition. The business became known as Frankfort Distilleries and Four Roses was their leading brand.

In 1943 Seagrams purchased Frankfort Distilleries and Straight Bourbons were phased out in favor of blended whiskey in the United States, but Four Roses remained a straight Bourbon in the overseas markets of Europe and Japan. In 2002 Seagrams was sold to Diageo who sold the Four Roses Brand and distillery to Kirin. Kirin immediately discontinued the Four Roses blended whiskey and brought the Bourbon back to the American market as well as the overseas markets.

Four Roses is unique among bourbon distilleries in that it uses two different mash bills (one high corn, one high rye) and five different yeast strains. By combining each of the two mash bills with each of the five yeast strains, Four Roses has 10 different bourbon recipes. They are used in different proportions in the different Four Roses expressions.

Mash bills:

E – 75% corn, 20% rye, 5% malted barley
B – 60% corn, 35% rye, 5% malted barley

Yeast Strain Codes:

- V – Light Fruitiness
- K – Slightly Spicy
- O – Rich Fruitiness
- F – Herbal
- Q – Floral

The four letter recipes are OBSV, OBSK, OBSQ, OBSO, OBSF and OESV, OESK, OESQ, OESO, OESF. The First O in each stands for the Lawrenceburg distillery, once known as Old Prentice. At one time this was one of five distilleries owned by Seagrams, which is how Four Roses wound up with five yeast strains. The S indicates that Four Roses is a straight bourbon.

In our notes we indicate which recipes are used in each bourbon. Finally, it is also of interest that Four Roses is unique among the bourbon distilleries in that it uses one story warehouses. Apparently, the ten recipes introduce enough variables to the bourbons without having to worry about aging at different rates at different warehouse levels!

In addition to the bourbons for which we have provided tasting notes, the distillery does annual, dated limited releases of specially selected Small Batch and Single Barrel bottlings.

Four Roses

Proof: 80
Age: NAS
Type: Straight Bourbon
Style: Traditional
Mash Bill: All 10 recipes
Color: Dark Straw
Price: $

Mike

Nose: Very floral (roses?), caramel and vanilla with light fruit. Just a hint of oak and spice.
Taste: Sweet with apples, pears and vanilla. Not hot at all. Nice mouthfeel.
Finish: Starts sweet but gets dry and peppery. Long lasting. One of the best finishes of any 80 proof bourbon.

Susan

Nose: Floral, light vanilla/caramel with underlying cinnamon notes.
Taste: Sweet vanilla with pear. Light and smooth on the tongue.
Finish: Starts sweet and dries towards the end. Lingering. No burn whatsoever.

Notes

A lot of presence for an 80 proof bourbon. Excellent to have on hand for making any of the classic bourbon cocktails.

My Score

Notes: ______________________________

____

____

Appearance (1-5) ____ Nose (1-25) ____
Taste (1-25) ____ Finish (1-25) ____
Complexity (1-10) ____ Overall Impression (1-10) ____

SCORE ____

Four Roses Elliott's Select

Proof: 106
Age: 14 years
Type: Straight
Style: Single Barrel, Barrel Proof, Extra Aged
Mash Bill: OESK
Color: Light Amber
Price: $$$

Mike

Nose: Crème brulee and fruit –ripe apples and berries – hint of spice.
Taste: Sweet berry fruit, vanilla and caramel with some oak.
Finish: Long and fruity with well-balanced oak.

Susan

Nose: Cinnamon, ripe berries and cherries with crème brulee. As it opens, cashews and pistachios emerge.
Taste: Fruit, caramel, vanilla and some milk chocolate.
Finish: Long, smooth, warm. A true Kentucky hug.

Notes

Brent Elliott's first signature release for Four Roses after succeeding Jim Rutledge as Master Distiller. This was Barrel 53-2Q.

My Score

Notes: ____________________________________

____

____

Appearance (1-5) ____ Nose (1-25) ____
Taste (1-25) ____ Finish (1-25) ____
Complexity (1-10) ____ Overall Impression (1-10) ____
SCORE ____

Four Roses Single Barrel

Proof: 100
Age: NAS
Type: Straight Bourbon
Style: Single Barrel
Mash Bill: OBSV
Color: Amber
Price: $$$

Mike

Nose: Sweet fruit. Berries and cherries with vanilla and caramel and a little spice.
Taste: Sweet berries and pears with some berries and spices. No burn and this is a 100 proof bourbon.
Finish: Long and sweet. Dry oak quickly becomes caramel and vanilla.

Susan

Nose: Vanilla, apple, pear with touch of cocoa and light spices.
Taste: Sweet vanilla and caramel with cinnamon sprinkled pear. Creamy mouthfeel.
Finish: Stays sweet to the very end of a long finish. Lingering cinnamon.

Notes

Complex and sophisticated, this still a bargain at its price. Another bourbon to savor post-dinner.

My Score

Notes: ____

____

____

Appearance (1-5) ____ Nose (1-25) ____
Taste (1-25) ____ Finish (1-25) ____
Complexity (1-10) ____ Overall Impression (1-10) ____

SCORE ____

Four Roses Single Barrel Private Selection

Proof: 116.6
Age: 10 years, 7 months
Type: Straight Bourbon
Style: Single Barrel, Barrel Proof, Extra Aged
Mash Bill: OESF
Color: Amber
Price: $$$

Mike

Nose: Very floral and minty – some caramel and tobacco/leather notes. Water brings out some honeysuckle.
Taste: Mint, caramel, apple fruit and tobacco. Water tames the mint and brings out fruit and liquorice.
Finish: Long and minty. Water makes the finish a bit sweeter with fruit and mint.

Susan

Nose: Spicy caramel, figs, dates, sweet tobacco. Water amplifies floral notes.
Taste: Add water to release sweet caramel, mint, with spicy oak and ripe figs.
Finish: Long, smooth, spicy with sweet saddle leather and fruit. Luscious.

Notes

Barrel proof single barrel selections are made by restaurants, bars, retailers, and bourbon clubs. They usually come with exact age statements. Four Roses enthusiasts will want to find all ten recipes. Addition of water reveals full complexity.

My Score

Notes: ______________________________________

____

____

Appearance (1-5) ____ Nose (1-25) ____
Taste (1-25) ____ Finish (1-25) ____
Complexity (1-10) ____ Overall Impression (1-10) ____

SCORE ____

Four Roses Small Batch

Proof: 90
Age: NAS
Type: Straight Bourbon
Style: Small Batch
Mash Bill: OBSO, OBSK, OESO, OESK
Color: Light Amber
Price: $$

Mike

Nose: Lots of apples, pears and flowers with vanilla and caramel and just a bit of oak.
Taste: Sweet fruit – pears. Vanilla and caramel with oaky spiciness. Nice mouthfeel – almost chewy.
Finish: Long with sweet fruit. Ends with caramel and just a touch of oak.

Susan

Nose: Caramel and vanilla with fresh pears and apples. Has oaky undertones.
Taste: Dark berry fruit, spices with vanilla and caramel creamy on the tongue.
Finish: Long, rich finish ending in light oak and a sprinkling of black pepper.

Notes

A fine sipping bourbon. Take it neat or with a splash of water.

My Score

Notes: ______________________________

____

____

Appearance (1-5) ____ Nose (1-25) ____
Taste (1-25) ____ Finish (1-25) ____
Complexity (1-10) ____ Overall Impression (1-10) ____

SCORE ____

Four Roses Small Batch Select

Proof: 104
Age: Minimum 6 years
Type: Straight
Style: Small Batch, Non-Chill Filtered
Mash Bill: OESV, OBSV, OBSK, OESK, OBSF, OESF
Color: Straw
Price: $$$

Mike

Nose: Very floral. Honeysuckle and roses with apples, pears, and old leather.
Taste: Very fruity with apples, pears, and a little berry fruit. Balanced by caramel and oak with a hint of baking spice.
Finish: Long and dry with oak and spice and a little lingering apple.

Susan

Nose: Caramel, butterscotch, nut bread, and a faint trace of dried fruit and leather.
Taste: Lots of creamy caramel and vanilla with walnuts, honey, cherries, and cinnamon.
Finish: Long and complex with dried fruit, nuts, and spices.

Notes

Uses both mash bills and three of the five yeast strains. Add water to create a fruit bomb.

My Score

Notes: ____________________________________

____

____

Appearance (1-5) ____ Nose (1-25) ____
Taste (1-25) ____ Finish (1-25) ____
Complexity (1-10) ____ Overall Impression (1-10) ____
SCORE ____

Four Roses Super Premium Platinum Label

Proof: 86
Age: 80% at least 8 years/20% 10 years
Type: Straight
Style: Export only to Japan
Mash Bill: OBSK, OESK, OBSV, OESV
Color: Dark Straw
Price: $$

Mike

Nose: Very floral – roses and apple blossom. A little anise/licorice, spice and oak.
Taste: Peppery spices and caramel. Vanilla accompanies by oak and anise.
Finish: Very long and dry as the oak tannins and and peppery spice quickly hide the sweetness.

Susan

Nose: Honey, orange blossom, toffee with prominent oak.
Taste: Sweet spices, rich vanilla/caramel. There's a touch of pear.
Finish: Long and smooth with sweet licorice at the very end.

Notes

Complex and assertive for such a moderately strong proof. Pleasant neat, but also a good choice for cocktail mixing.

My Score

Notes: ____

____

____

Appendix (1-5) ____ Nose (1-25) ____
Taste (1-25) ____ Finish (1-25) ____
Complexity (1-10) ____ Overall Impression (1-10) ____
SCORE ____

Four Roses Additional Releases

Name: ____________________

Distillery: ____________________

Proof: ________ **Mash Bill:** ________

Age: ________ ________

Type: ________ **Color:** ________

Style: ________ **Price:** ________

My Score

Notes: ____________________

____

____

Appearance (1-5) ____ Nose (1-25) ____

Taste (1-25) ____ Finish (1-25) ____

Complexity (1-10) ____ Overall Impression (1-10) ____

SCORE ____

Name: ____________________

Distillery: ____________________

Proof: ________ **Mash Bill:** ________

Age: ________ ________

Type: ________ **Color:** ________

Style: ________ **Price:** ________

My Score

Notes: ____________________

____

____

Appearance (1-5) ____ Nose (1-25) ____

Taste (1-25) ____ Finish (1-25) ____

Complexity (1-10) ____ Overall Impression (1-10) ____

SCORE ____

Green River

www.greenriverdistilling.com

Bourbon Distilleries:

- Green River Distillery, Owensboro, Ky.

When it was built in 1885, it was called McCullough before becoming Green River and one of many distilleries in the Owensboro area. Just before Prohibition took effect, the original distillery burned. It was rebuilt in 1938 and operated as the Medley Distillery until its then owner, United Distillers, shut it down in 1992. It sat vacant for many years until it was acquired by Terressentia. It was renamed O.Z. Tyler, but in 2020 it returned to its original name. Like Bardstown Bourbon Company, it also engages in a lot of contract distilling, as well as releasing its own brands.

Bradshaw Bourbon

Proof: 103.8
Age: 2 years
Type: Straight
Style: Small Batch
Mash Bill: 70% corn, 21% rye, 9% malted barley
Color: Light Straw
Price: $$

Mike

Nose: Vanilla, hint of apples and spice, and a little oak.
Taste: Vanilla, with a little green apple and pepper spice and oak.
Finish: Medium long with pepper and a hint of oak.

Susan

Nose: Maple candy and pecans. Praline. Some brown sugar and oak, too.
Taste: Caramel apple and some berries. Tannins kick in quickly.
Finish: Short and tannic, but not bitter.

Notes

NFL quarterback great Terry Bradshaw is behind the brand made for his company at Green River. Bottled at 51.9% abv as a nod to his passing completion rate.

My Score

Notes: ________________________________

____

____

Appearance (1-5) ____ Nose (1-25) ____
Taste (1-25) ____ Finish (1-25) ____
Complexity (1-10) ____ Overall Impression (1-10) ____
SCORE ____

Cumberland Falls

Proof: 86
Age: 4 years
Type: Straight
Style: Small Batch
Mash Bill: 70% corn, 21% rye, 9% malted barley*
Color: Straw
Price: $$

Mike

Nose: Corn, vanilla, apricots, white pepper, oak.
Taste: Corn, vanilla, apricots, white pepper, oak.
Finish: Medium long with oak and pepper.

Susan

Nose: Vanilla, milk chocolate, pears, sweet mint.
Taste: Vanilla pudding with some milk chocolate, baking spice, and new leather.
Finish: Dry and peppery.

Notes

*Not stated on the label, but this is probably the mash bill given that it's the one used in other Green River bourbons.

My Score

Notes: ______________________________
____
____

Appearance (1-5) ____ Nose (1-25) ____
Taste (1-25) ____ Finish (1-25) ____
Complexity (1-10) ____ Overall Impression (1-10) ____

SCORE ____

Green River

Proof: 90
Age: NAS (~5 years)
Type: Straight
Style: Traditional
Mash Bill: 70% corn, 21% winter rye, 9% 2-row barley
Color: Straw
Price: $$

Mike

Nose: Vanilla, a little corn with apricot and peaches. Oak.
Taste: Vanilla, corn, apricots with cardamom, allspice, and oak.
Finish: Long with oak and baking spices.

Susan

Nose: Brown sugar, candies orange peel, light dusting of cinnamon.
Taste: Brown sugar, fruit, and spice outmatched by oaky tannins.
Finish: Lingering heat.

Notes

Revival of a pre-Prohibition brand with the wonderful tag line," The Whiskey with No Regrets."

My Score

Notes: ______________________________

____

____

Appearance (1-5) ___ Nose (1-25) ___
Taste (1-25) ___ Finish (1-25) ___
Complexity (1-10) ___ Overall Impression (1-10) ___
SCORE ___

Kentucky 10

Proof: 90
Age: 2 years
Type: Straight
Style: Wheated
Mash Bill: Not available, but contains wheat.
Color: Straw
Price: $$

Mike

Nose: Corn, vanilla, a hint of floral notes, oak.
Taste: Corn, vanilla, a little sweet pear fruit, oak, and spice.
Finish: Medium long with oak and spice – clove and allspice.

Susan

Nose: Light brown sugar, a little green apple peel, a fair amount of oak.
Taste: Brown sugar is joined by some faintly floral notes and a good deal of pepper.
Finish: Strats with pepper flakes, but then mellows to brown sugar and a touch of fruit.

Notes

Nice for a two-year-old. Produced by Green River Distilling Company, Owensboro, Kentucky.

My Score

Notes: ______________________________

____

____

Appearance (1-5) ____ Nose (1-25) ____
Taste (1-25) ____ Finish (1-25) ____
Complexity (1-10) ____ Overall Impression (1-10) ____

SCORE ____

Yellow Banks Straight Bourbon

Proof: 92
Age: 2 years
Type: Straight
Style: Traditional
Mash Bill: 70% corn, 21% rye, 9% malted barley
Color: Light Straw
Price: $$

Mike

Nose: Light with corn, vanilla, pears, and oak.
Taste: Corn, vanilla, pears, oak, and leather.
Finish: Medium long. Pear lingers with some oak and spice.

Susan

Nose: Caramel and nutmeg with light oak.
Taste: Echoes the nose, plus some fresh peaches.
Finish: Long with oak, but not bitter.

Notes

Rich mouthfeel for a young bourbon.

My Score

Notes: ______________________________

____

____

Appearance (1-5) ____ Nose (1-25) ____
Taste (1-25) ____ Finish (1-25) ____
Complexity (1-10) ____ Overall Impression (1-10) ____
SCORE ____

Green River Additional Releases

Name: ______________________________

Distillery: ______________________________

Proof: ____________ **Mash Bill:** ____________
Age: ____________ ____________
Type: ____________ **Color:** ____________
Style: ____________ **Price:** ____________

My Score

Notes: ______________________________
____
____

Appearance (1-5) ___ Nose (1-25) ___
Taste (1-25) ___ Finish (1-25) ___
Complexity (1-10) ___ Overall Impression (1-10) ___

SCORE ___

Name: ______________________________

Distillery: ______________________________

Proof: ____________ **Mash Bill:** ____________
Age: ____________ ____________
Type: ____________ **Color:** ____________
Style: ____________ **Price:** ____________

My Score

Notes: ______________________________
____
____

Appearance (1-5) ___ Nose (1-25) ___
Taste (1-25) ___ Finish (1-25) ___
Complexity (1-10) ___ Overall Impression (1-10) ___

SCORE ___

Heaven Hill

www.heavenhilldistillery.com

Bourbon Distilleries:

- Bernheim Distillery, Louisville, Ky.
- Evan Williams Bourbon Experience, Louisville, Ky.

The five Shapira brothers, – Gary, Mose, George, Edward and David – decided to go into the distilling business after the repeal of Prohibition. They purchased the land once owned by William Heavenhill near Bardstown and built a distillery. The distillery opened in 1935 and the brothers hired a Beam to make their whiskey. It is still a family owned business with Edward's son Max at the helm. In 1996 a fire destroyed the distillery in Bardstown and the company purchased the Bernheim distillery in Louisville from Diageo. In 2014, Heaven Hill opened a small craft distillery and tour on Main Street in Louisville where they make one barrel of whiskey a day.

Heaven Hill has the second-largest stock of aging spirits in Kentucky, after Jim Beam. It contract distills for a number of independent bottlers.

Heaven Hill has two mash bills for its bourbons, one using rye as the flavoring grain. The other uses wheat. Limited release Parker's Heritage whiskeys have a variety of mash bills.

Cabin Still

Proof:	80
Age:	36 months
Type:	Straight Bourbon
Style:	Traditional
Mash Bill:	78% corn, 10% rye, 12% malted barley
Color:	Light Straw
Price:	$

Mike

Nose: Corn and vanilla. A hint of oak. Not complex.
Taste: Corn and vanilla with some oak tannins.
Finish: Short, but gets some oak and pepper toward the end.

Susan

Nose: Some mint, sweet corn, caramel.
Taste: Sweet on the tongue. Again, mint with a little sweet spice.
Finish: Medium long. Stays warm and sweet.

Notes

Considering that you can find this bourbon for under $10, it is well worth trying. Nice, light sip that will be a good "summer bourbon" with an ice cube.

My Score

Notes: ____

____

____

Appearance (1-5)	____	Nose (1-25)	____
Taste (1-25)	____	Finish (1-25)	____
Complexity (1-10)	____	Overall Impression (1-10)	____

SCORE ____

Elijah Craig Single Barrel

Proof: 132.4
Age: 12 years
Type: Straight Bourbon
Style: Single Barrel, Barrel Proof, Extra Aged
Mash Bill: 78% corn, 10% rye, 12% malted barley
Color: Rich Amber
Price: $$

Mike

Nose: Caramel with a hint of milk chocolate, ripe pears and apples, fine leather, sweet spices.
Taste: Sweet fruit – apricots, pears, and berries caramel and vanilla with nutmeg. Very rich mouthfeel with little burn.
Finish: Long and spicy with sweet oak.

Susan

Nose: Ripe figs, crème brûlée, roasted corn, and alcohol of its high proof.
Taste: Add water! It reveals ripe apple, pears and roasted corn. Multi-layered.
Finish: Long, sweet, and smooth.

Notes

This may prove difficult to find. Heaven Hill keeps changing age and availability of its Elijah Craig expressions. By the way, all the stories about Baptist minister Elijah Craig having been the first to put whiskey in charred oak barrels are fun, but mythological.

My Score

Notes: ______________________________________
____
____

Appearance (1-5) ____ Nose (1-25) ____
Taste (1-25) ____ Finish (1-25) ____
Complexity (1-10) ____ Overall Impression (1-10) ____
SCORE ____

Elijah Craig Small Batch

Proof: 94
Age: NAS, 8-12 year batched barrels
Type: Straight
Style: Small Batch
Mash Bill: 78% corn, 10% rye, 12% malted barley
Color: Dark Straw
Price: $$$

Mike

Nose: Vanilla, peaches, baking spice, oak.
Taste: Vanilla, peach and apricot, allspice, white pepper, oak.
Finish: Medium long with oak and pepper spice.

Susan

Nose: Caramel candy, whiff of mint, with peaches and a dash of cinnamon.
Taste: Vanilla pudding drizzled with milk chocolate, pitted fruit, and cinnamon.
Finish: Warm and beautifully balanced.

Notes

Truly versatile. Sip neat or use in a spirits forward cocktail.

My Score

Notes: ______________________________

____

____

Appearance (1-5) ____ Nose (1-25) ____
Taste (1-25) ____ Finish (1-25) ____
Complexity (1-10) ____ Overall Impression (1-10) ____

SCORE ____

Elijah Craig 12-Year-Old

Proof: 94
Age: 12 years
Type: Straight Bourbon
Style: Small Batch, Extra Aged
Mash Bill: 78% corn, 10% rye, 12% malted barley
Color: Light Amber
Price: $$$

Mike

Nose: Crème brûlée, apricots, and oak. Fine tobacco and leather with a hint of spice.
Taste: Apricots and caramel with some oak wood and spice.
Finish: Long. Starts sweet but finishes with oak and a hint of rotten peach.

Susan

Nose: Ripe peaches, roasted corn, caramel candy with a touch of wood smoke.
Taste: Pitted fruits, soft vanilla and buttered popcorn. Buttery mouthfeel.
Finish: Long and warm. Very smooth.

Notes

Not especially high proof, but just a few drops of water will amplify the fruit character. Very nice neat, too. Excellent for the price.

My Score

Notes: ______________________________

____

____

Appearance (1-5) ___ Nose (1-25) ___
Taste (1-25) ___ Finish (1-25) ___
Complexity (1-10) ___ Overall Impression (1-10) ___

SCORE ___

Elijah Craig 18-Year-Old

Proof: 86
Age: 18 years
Type: Straight
Style: Single Barrel, Extra Aged
Mash Bill: 78% corn, 10% rye, 12% malted barley
Color: Straw
Price: $$$

Mike

Nose: Caramel, vanilla, and old leather.
Taste: Caramel, vanilla, fruit and spice.
Finish: Oak with a hint of rotten peach.

Susan

Nose: Vanilla, pears, hard candy.
Taste: Vanilla custard, almonds, marzipan, with some orchard fruit.
Finish: Long, smooth and sweet.

Notes

This was barrel number 4146. A few short years ago you could buy this bourbon for about $45 a bottle. No more!

My Score

Notes: ______________________________

____

____

Appearance (1-5) ____ Nose (1-25) ____
Taste (1-25) ____ Finish (1-25) ____
Complexity (1-10) ____ Overall Impression (1-10) ____

SCORE ____

Elijah Craig 21-Year-Old

Proof: 90
Age: NAS
Type: Straight Bourbon
Style: Single Barrel, Extra Aged
Mash Bill: 78% corn, 10% rye, 12% malted barley
Color: Dark Straw
Price: $$$

Mike

Nose: Vanilla and peaches with baking spice. Corn husk, nuts, hay loft.
Taste: Corn, vanilla, and spice.
Finish: Spice with a hint of rotten peaches.

Susan

Nose: Corn husk, nuts, hayloft.
Taste: Very herbal. Not unpleasant, just different. There's a bit of chocolate, too
Finish: Surprisingly sweet finish with a coconut note.

Notes

This was from barrel number 18. A complex sip; we obviously had very different impressions of this bourbon!

My Score

Notes: ________________________________

____

____

Appearance (1-5) ____ Nose (1-25) ____
Taste (1-25) ____ Finish (1-25) ____
Complexity (1-10) ____ Overall Impression (1-10) ____

SCORE ____

Elijah Craig 23-Year-Old

Proof: 90
Age: 23 years
Type: Straight Bourbon
Style: Single Barrel, Extra Aged
Mash Bill: 78% corn, 10% rye, 12% malted barley
Color: Dark Amber/Bronze
Price: $$$

Mike

Nose: Crème brûlée, apricot, oak, fine leather and little tobacco.
Taste: Sweet fruit – peaches and apricots – caramel and vanilla. A bit of nutmeg and fine tobacco.
Finish: Very simple and short. The oak lasts longest, giving it a bit of spice.

Susan

Nose: Honey, dates, fruit cobbler, and sweet oaky wood.
Taste: Honey, caramel apple, a flash of coconut Leather and tobacco balanced against fruit and spices.
Finish: Long and lingering with warm fruit. A masterpiece.

Notes

This was Barrel 26, filled on 26 February, 1990. The retail price hovers around $200, so worthy of the best glassware and the finest cigar. Let it sit in the glass for a bit for all the layers to develop.

My Score

Notes: ____________________________________

____

____

Appearance (1-5) ____ Nose (1-25) ____
Taste (1-25) ____ Finish (1-25) ____
Complexity (1-10) ____ Overall Impression (1-10) ____
SCORE ____

Evan Williams

Proof: 80
Age: NAS
Type: Straight Bourbon
Style: Traditional
Mash Bill: 78% corn, 10% rye, 12% malted barley
Color: Dark Straw
Price: $

Mike

Nose: Corn and vanilla with some sweet spice and oak. Not overly complex.

Taste: Nice mouthfeel with no alcohol burn, but warmth in the throat. Sweet corn and vanilla with a bit of apple and spice.

Finish: Starts sweet, but quickly dries to a nice oak finish.

Susan

Nose: Sweet buttered corn and vanilla. Light.

Taste: Sweet and smooth with a pleasant buttery texture. Candied nuts and sweet spices on the palate.

Finish: Long, smooth, and sweet. No burn.

Notes

A very good everyday bourbon at an excellent price. Makes a good Old Fashioned.

My Score

Notes: ______________________________

____

____

Appearance (1-5) ____ Nose (1-25) ____
Taste (1-25) ____ Finish (1-25) ____
Complexity (1-10) ____ Overall Impression (1-10) ____

SCORE ____

Evan Williams Barrel Proof

Proof:	125
Age:	NAS
Type:	Straight Bourbon
Style:	Barrel Proof
Mash Bill:	78% corn, 10% rye, 12% malted barley
Color:	Amber
Price:	$$$

Mike

Nose: Corn and caramel with some fruit and spices.
Taste: Corn, vanilla, oak, and spice. Not hot, but warm in the throat.
Finish: Very nice. Starts oak dry but sweetens with caramel.

Susan

Nose: Vanilla, peaches, pistachios, new leather.
Taste: Very fruity entry. Then crème brulee and nuts.
Finish: Peppery at barrel proof, but water reveals caramel.

Notes

Definitely add a bit of water or let an ice cube melt to reveal the full complexity of this bourbon.

My Score

Notes: ______________________________________

____

____

Appearance (1-5)	____	Nose (1-25)	____
Taste (1-25)	____	Finish (1-25)	____
Complexity (1-10)	____	Overall Impression (1-10)	____

SCORE ____

Evan Williams Black Label

Proof: 86
Age: NAS
Type: Straight Bourbon
Style: Traditional
Mash Bill: 78% corn, 10% rye, 12% malted barley
Color: Dark Straw
Price: $

Mike

Nose: Honey, vanilla, corn. Some pear fruit.
Taste: Pears and caramel with sweet spices (nutmeg or allspice?).
Finish: Long. Starts sweet and dries to oak and pepper.

Susan

Nose: Sweet vanilla, dash of cinnamon, hint of cocoa.
Taste: Vanilla, ripe peach, sweet spices. Hint of sweet mint Rich mouthfeel.
Finish: Warm and quite smooth. No burn. Ends in sweet oak.

Notes

An outstanding bargain in bourbon. Nice with an ice cube. Excellent for mixing cocktails.

My Score

Notes: ______________________________

____

____

Appearance (1-5) ____ Nose (1-25) ____
Taste (1-25) ____ Finish (1-25) ____
Complexity (1-10) ____ Overall Impression (1-10) ____

SCORE ____

Evan Williams White Label

Proof:	100
Age:	NAS
Type:	Straight Bourbon
Style:	Bonded
Mash Bill:	78% corn, 10% rye, 12% malted barley
Color:	Dark Straw
Price:	$

Mike

Nose: Vanilla/caramel with some apple and apricot, cinnamon and nutmeg and a hint of oak.

Taste: Great mouthfeel, almost chewy with no burn. Apples, cinnamon, caramel, vanilla. Fine leather appears last.

Finish: Very nice with a bit of sweetness leading into some oak. Medium long.

Susan

Nose: Sweet caramel corn, cotton candy, vanilla ice cream, sweet spices.

Taste: Crème brûlée with candied almonds. Very smooth, sweet sip.

Finish: Sweetness lingers, but not cloying. Touch of oak at the end.

Notes

Very approachable bottled-in-bond bottle and an excellent price.

My Score

Notes: ______________________________________

____

____

Appearance (1-5) ____ Nose (1-25) ____

Taste (1-25) ____ Finish (1-25) ____

Complexity (1-10) ____ Overall Impression (1-10) ____

SCORE ____

Evan Williams Single Barrel 2004

Proof:	86.6
Age:	10 years
Type:	Straight Bourbon
Style:	Single Barrel, Extra Aged
Mash Bill:	78% corn, 10% rye, 12% malted barley
Color:	Light Amber
Price:	$$

Mike

Nose: Sweet fruit – apricots and berries with caramel and oak.

Taste: Caramel and apricots with some peppery spice, old leather, and oak.

Finish: Long and dry. Starts fruity and spicy but dries to nice oak finish.

Susan

Nose: Caramel, cocoa, cinnamon, and orchard fruits.

Taste: Cherry and caramel with a little ginger. Buttered corn.

Finish: Long and sweet, ending in sweet oak.

Notes

Another bargain-priced goody from Heaven Hill. Sip neat, but also excellent for cocktails. This was Barrel 591.

My Score

Notes: ____________________

____

____

Appearance (1-5)	___	Nose (1-25)	___
Taste (1-25)	___	Finish (1-25)	___
Complexity (1-10)	___	Overall Impression (1-10)	___

SCORE ___

Evan Williams 12-Year-Old

Proof: 101
Age: 12 years
Type: Straight
Style: Extra Aged
Mash Bill: 78% corn, 10% rye, 12% malted barley
Color: Amber/Copper
Price: $$$$

Mike

Nose: Caramel and apples with ginger and nutmeg, fine tobacco and leather. Water brings out the oak.
Taste: Caramel, apple, hint of chocolate. Nutmeg cinnamon with some tobacco. Water amplifies the chocolate.
Finish: Long, but not too long. Oak and some pepper. Water removes most of the pepper and enhances the oak.

Susan

Nose: Chocolate-covered caramel, cashews and ripe apples.
Taste: Honey, caramel, cocoa. Water reveals apples and more caramel.
Finish: Long, smooth and caramel sweet. Dries to sweet oak.

Notes

The retail price is more than $100. But this "Extra Smooth Sour Mash" delivers.

My Score

Notes: ____________________________________

____

____

Appearance (1-5) ____ Nose (1-25) ____
Taste (1-25) ____ Finish (1-25) ____
Complexity (1-10) ____ Overall Impression (1-10) ____

SCORE ____

Evan Williams 23-Year-Old

Proof: 107
Age: 23 years
Type: Straight Bourbon
Style: Extra Aged
Mash Bill: 78% corn, 10% rye, 12% malted barley
Color: Dark Amber
Price: $$$$

Mike

Nose: Caramel and chocolate with some fruit and spice. Fine tobacco and oak.

Taste: Not very hot, with some sweet caramel, oak and a hint of must or mold.

Finish: Starts sweet and dries nicely with a long oak finish.

Susan

Nose: Honey and molasses with notes of pears and sweet spices.

Taste: Closed at first, but water brings out pear, apple, honey. Complex, with a little cinnamon and leather.

Finish: Long and warm and buttery, with a pleasantly dry ending.

Notes

A little water lessens the mustiness detected by Mike. Limited annual release and one of the most expensive bourbons on the market.

My Score

Notes: ____________________________________

____

____

Appearance (1-5) ____ Nose (1-25) ____

Taste (1-25) ____ Finish (1-25) ____

Complexity (1-10) ____ Overall Impression (1-10) ____

SCORE ____

Evan Williams 1783

Proof: 86
Age: NAS
Type: Straight Bourbon
Style: Small Batch
Mash Bill: 78% corn, 10% rye, 12% malted barley
Color: Medium Dark Straw
Price: $

Mike

Nose: Very light. Corn and vanilla with a bit of spice.
Taste: A bit thin in the mouth – corn and vanilla with some apple fruit and pepper spice.
Finish: Short. A bit of corn sweetness but fades quickly with only a hint of oak.

Susan

Nose: Light caramel with herbal notes. A little burnt sugar and cocoa powder.
Taste: Nutty. Vanilla fades into corn and sweet spices.
Finish: Somewhat short, but a bright oak note at the end.

Notes

Evan Williams was a Louisville brick maker and distiller, who had one of the earliest commercial distilleries in Kentucky. Yes, the distilling license was granted in 1783.

My Score

Notes: ______________________________

____

____

Appearance (1-5) ____ Nose (1-25) ____
Taste (1-25) ____ Finish (1-25) ____
Complexity (1-10) ____ Overall Impression (1-10) ____

SCORE ____

Fighting Cock

Proof: 103
Age: 6 years
Type: Straight Bourbon
Style: Traditional
Mash Bill: 78% corn, 10% rye, 12% malted barley
Color: Light Amber
Price: $

Mike

Nose: Vanilla and oak with leather and sweet spice. A little honeysuckle.
Taste: Thick, chewy mouthfeel. Ripe apple and pepper spice with fine tobacco.
Finish: This is where the fight comes in, with boatloads of spicy pepper and oak.

Susan

Nose: Big fruit and caramel nose with hazelnut and vanilla bean.
Taste: Nutty caramel and chocolate. Truly a Snickers Bar in a glass.
Finish: Buttered popcorn that rolls across the tongue and finishes in pepper and oak.

Notes

A very well-balanced bourbon and an excellent value with age and proof. "I like it. Too bad about the name," says Susan.

My Score

Notes: ____

____

____

Appearance (1-5) ____ Nose (1-25) ____
Taste (1-25) ____ Finish (1-25) ____
Complexity (1-10) ____ Overall Impression (1-10) ____

SCORE ____

Five Brothers Bourbon

Proof:	90
Age:	Batched 5- to 9-year bourbons
Type:	Straight
Style:	Small Batch
Mash Bill:	70% corn, 20% rye, 10% malted barley
Color:	Straw
Price:	$$$

Mike

Nose: Caramel, apples, a little cinnamon, oak, and tobacco.

Taste: Caramel apples, cinnamon, nutmeg, ginger, oak, and tobacco.

Finish: Long with oak and spices.

Susan

Nose: Roasted corn, some apricot, caramel, and vanilla.

Taste: Corn on the cob, sweet mint, new leather, and pipe tobacco.

Finish: Dries to herbal nuttiness.

Notes

Released in 2021 to commemorate the five Shapira brothers who founded the company in 1935. Available in the gift shop at the Heaven Hill Bourbon Experience in Bardstown.

My Score

Notes: ______________________________

____

____

Appearance (1-5) ____ Nose (1-25) ____

Taste (1-25) ____ Finish (1-25) ____

Complexity (1-10) ____ Overall Impression (1-10) ____

SCORE ____

Heaven Hill Grain to Glass Straight Bourbon

Proof:	107
Age:	7 years
Type:	Straight
Style:	Traditional
Mash Bill:	52% corn, 35% rye, 15% malted barley
Color:	Straw
Price:	$$$$

Mike

Nose: Vanilla, fruit, cardamom, oak.
Taste: Vanilla, peaches, apricots, cardamom, and oak.
Finish: Long with oak and cardamom that turns to pepper at the end.

Susan

Nose: Apples, pears, barn loft, soft pepper.
Taste: Apples and pears with oak and pepper. Touch of honey and brown sugar.
Finish: Long, drying to spice and light oak.

Notes

Annual limited release. The corn is a specially developed hybrid (Beck's 6158) for this series.

My Score

Notes: ______________________________

____

____

Appearance (1-5) ____ Nose (1-25) ____

Taste (1-25) ____ Finish (1-25) ____

Complexity (1-10) ____ Overall Impression (1-10) ____

SCORE ____

Heaven Hill Grain to Glass Wheated

Proof: 121
Age: 7 years
Type: Straight
Style: Wheated
Mash Bill: 52% corn, 35% wheat, 15% malted barley
Color: Light Amber
Price: $$$$

Mike

Nose: Vanilla, apples, hazelnut, pepper, and oak.
Taste: Vanilla, apples, black pepper, and oak.
Finish: Long with oak and pepper.

Susan

Nose: Nutty. Some almond and citrus with creme brulee and nutmeg.
Taste: Vanilla wafers, almonds, some cinnamon and sorghum.
Finish: Long and spicy.

Notes

Annual limited release. The corn is a specially developed hybrid (Beck's 6158) for this series. There is a rye in the series, too.

My Score

Notes: ____________________________________
____
____

Appearance (1-5) ____ Nose (1-25) ____
Taste (1-25) ____ Finish (1-25) ____
Complexity (1-10) ____ Overall Impression (1-10) ____
SCORE ____

Heaven Hill Old Style Bourbon Green Label

Proof: 80
Age: 36 months
Type: Straight Bourbon
Style: Traditional
Mash Bill: 78% corn, 10% rye, 12% malted barley
Color: Light Straw
Price: $

Mike

Nose: Very light. Corn and vanilla with a hint of spiced oak.
Taste: Corn and vanilla with some pepper spice. Light, but not watery.
Finish: Medium long with some oak spiciness.

Susan

Nose: Sweet vanilla, light spices, ripe apple. Surprisingly present for age and proof.
Taste: Vanilla with a touch of oak. Not much fruit or spice on the palate.
Finish: Medium. Starts sweet and fades to warm, black pepper.

Notes

Lightweight and young, but this could be a very nice introductory bourbon for someone afraid that whiskey is "harsh." Very pleasant, if simple.

My Score

Notes: ______________________________

____

____

Appearance (1-5) ____ Nose (1-25) ____
Taste (1-25) ____ Finish (1-25) ____
Complexity (1-10) ____ Overall Impression (1-10) ____

SCORE ____

Heaven Hill Select Stock

Proof: 128
Age: 8 years
Type: Special Finish
Style: Single Barrel, Barrel Proof
Mash Bill: 78% corn, 10% rye, 12% malted barley
Color: Light Amber
Price: $$$$

Mike

Nose: Crème brulee, fruit and hint of oak.
Taste: Caramel and raisins with some pepper spice.
Finish: Oak and alcohol. Water helps and makes the finish less astringent.

Susan

Nose: Vanilla and pears with dash of cinnamon.
Taste: Crème brulee, ripe pears, almonds.
Finish: Hot at barrel proof. Much sweeter with water.

Notes

Definitely add water! Bit warm in the mouth. Dash of water takes away the burn and leaves the flavors intact. Bourbon was finished in used cognac barrels. This was barrel #44823. Distilled 9-24-03. Bottled 5-5-14.

My Score

Notes: ______________________________

____

____

Appearance (1-5) ____ Nose (1-25) ____
Taste (1-25) ____ Finish (1-25) ____
Complexity (1-10) ____ Overall Impression (1-10) ____

SCORE ____

Henry McKenna

Proof: 80
Age: NAS
Type: Straight Bourbon
Style: Traditional
Mash Bill: 78% corn, 10% rye, 12% malted barley
Color: Straw
Price: $

Mike

Nose: Very light. Vanilla and candy corn with a hint of oak.
Taste: Bit thin and watery. Vanilla and corn, with some pepper spice. Little heat.
Finish: Starts sweet and dries nicely with a long oak finish.

Susan

Nose: Light. Some ripe peach and a little vanilla and sweet corn.
Taste: Candy corn and peppery spice. Not complex but pleasant.
Finish: Long and warm and buttery, with a pleasantly dry ending.

Notes

Nice, inexpensive whiskey for a mixed drink such as bourbon and ginger ale.

My Score

Notes: ____

____

____

Appearance (1-5) ____ Nose (1-25) ____
Taste (1-25) ____ Finish (1-25) ____
Complexity (1-10) ____ Overall Impression (1-10) ____

SCORE ____

Henry McKenna Bottled-in-Bond

Proof: 100
Age: 10 years
Type: Straight Bourbon
Style: Bonded, Single Barrel
Mash Bill: 78% corn, 10% rye, 12% malted barley
Color: Dark Straw
Price: $$

Mike

Nose: Caramel and vanilla up front. Old leather, ripe peach or apricot, and oak.
Taste: Very soft on the tongue with fruit and spice – nutmeg. A bit of honey and oak toward the end.
Finish: Starts sweet and dries nicely with a long oak finish.

Susan

Nose: Sweet, floral, with vanilla and caramel candy.
Taste: Caramel apple with sweet vanilla and powdered sugar. Drop of water brings out cherries and chocolate.
Finish: Long, smooth, and warm. No burn.

Notes

This was Barrel No. 1210. Don't be afraid to add a little water to bring out more fruit and spice.

My Score

Notes: ________________________________

____

____

Appearance (1-5) ____ Nose (1-25) ____
Taste (1-25) ____ Finish (1-25) ____
Complexity (1-10) ____ Overall Impression (1-10) ____
SCORE ____

J.T.S. Brown Bottled-in-Bond

Proof: 100
Age: At least 4 years
Type: Straight Bourbon
Style: Bottled-in-Bond
Mash Bill: 78% corn, 10% rye, 12% malted barley
Color: Light Straw
Price: $

Mike

Nose: Vanilla and corn with a hint of oak.
Taste: Corn and vanilla with a hint of honey.
Finish: Dry and spicy with oak and pepper.

Susan

Nose: Vanilla, banana, nutmeg.
Taste: Caramel entry, but quite spicy on the tongue.
Finish: Short, peppery.

Notes

Perhaps a bit lightweight for a bottled-in-bond. But, very interesting to compare to other B.I.B. brands.

My Score

Notes: ____________________________________

____

____

Appearance (1-5) ____ Nose (1-25) ____
Taste (1-25) ____ Finish (1-25) ____
Complexity (1-10) ____ Overall Impression (1-10) ____
SCORE ____

Larceny

Proof: 92
Age: NAS
Type: Straight Bourbon
Style: Wheated
Mash Bill: 68% corn, 20% wheat, 10% malted barley
Color: Straw
Price: $

Mike

Nose: Not complex. Vanilla/caramel with some apples and pears.
Taste: Sweet with vanilla/caramel and apple and honey. Hint of pecan and sweet spices.
Finish: Starts sweet with caramel and dries out with some nice oak tannins. Long and lightly dry.

Susan

Nose: Touch of honey under the vanilla and caramel.
Taste: Sweet vanilla custard. Some marzipan. A little nutmeg and cinnamon.
Finish: Medium long and smooth with a touch of dry oak at the very end.

Notes

Good value in a wheated bourbon.

My Score

Notes: ____________________________________

____

____

Appearance (1-5) ____ Nose (1-25) ____
Taste (1-25) ____ Finish (1-25) ____
Complexity (1-10) ____ Overall Impression (1-10) ____
SCORE ____

Larceny Barrel Proof

Proof: 123.8
Age: NAS
Type: Straight
Style: Barrel Proof, Wheated, Non-Chill Filtered
Mash Bill: 68% corn, 20% wheat, 12% malted barley
Color: Light Amber
Price: $$$

Mike

Nose: Caramel, apples, nutmeg, allspice, oak.
Taste: Caramel, apples, nutmeg, cinnamon, oak.
Finish: Long and dry with oak and spice.

Susan

Nose: Vanilla pudding, caramel, plums, cinnamon, clove.
Taste: Caramel, berries, baking spices. Rich mouthfeel.
Finish: Long and warm with caramel and sweet oak.

Notes

Batched from a mingling of 6 to 8 year old bourbons.

My Score

Notes: ______________________________

____

____

Appearance (1-5) ____ Nose (1-25) ____
Taste (1-25) ____ Finish (1-25) ____
Complexity (1-10) ____ Overall Impression (1-10) ____

SCORE ____

Old Fitzgerald Bottled-in-Bond

Proof: 100
Age: NAS
Type: Straight Bourbon
Style: Wheated
Mash Bill: Proprietary
Color: Dark Straw
Price: $$$$

Mike

Nose: Vanilla and hazelnuts with some apple fruit and a whisper of oak.
Taste: Vanilla and apples with some oak and pecan. A bit of pepper spice toward the end. Nice warmth in the throat, but not hot.
Finish: Long and spicy with a nice oak dryness.

Susan

Nose: Honey, caramel, whiff of sweet spice. A hint of cocoa. Very smooth.
Taste: Honey, candied apples, vanilla, and buttered corn. A little water brings out cinnamon.
Finish: Lingering with light oak and pepper.

Notes

Elegant bonded bourbon that becomes fruitier and sweeter with a little bit of water. But don't add too much.

My Score

Notes: ____________________________________

____

____

Appearance (1-5) ____ Nose (1-25) ____
Taste (1-25) ____ Finish (1-25) ____
Complexity (1-10) ____ Overall Impression (1-10) ____
SCORE ____

Old Fitzgerald Bottled-in-Bond 2022

Proof: 100
Age: 19 years
Type: Straight
Style: Bonded, Wheated
Mash Bill: 68% corn, 20% wheat, 12% malted barley
Color: Light Straw
Price: $$$$

Mike

Nose: Light – vanilla, apples, hint of spice and oak.
Taste: Vanilla, apples, pepper, and oak.
Finish: Medium long with oak, pepper, and lingering apple.

Susan

Nose: Light, but some vanilla and ripe apple. A little baking spice.
Taste: Corn, caramel, pepper, and pitted fruit.
Finish: Long and peppery.

Notes

Batched from a mingling of 6 to 8 year old bourbons. Fall 2022 release.

My Score

Notes: ______________________________

____

____

Appearance (1-5) ____ Nose (1-25) ____
Taste (1-25) ____ Finish (1-25) ____
Complexity (1-10) ____ Overall Impression (1-10) ____

SCORE ____

Old Fitzgerald Bottled-in-Bond 2024

Proof: 100
Age: 13 years
Type: Straight
Style: Bottled-in-Bond
Mash Bill: 68% corn, 20% wheat, 12% malted barley
Color: Dark Straw
Price: $$$$

Mike

Nose: Vanilla and caramel, apples, pepper, and oak.
Taste: Vanilla and caramel, apples, pepper, and oak.
Finish: Long and dry with oak and pepper.

Susan

Nose: Toasting wheat bread, new leather, light cherries, cinnamon.
Taste: Vanilla wafers, dark cherries, a bit of tobacco leaf, and black pepper.
Finish: Warm and dry with oak and pepper.

Notes

Complex and beautifully balanced. Annual limited decanter release. You will pay big bucks on the secondary market.

My Score

Notes: ____________________________________

____

____

Appearance (1-5) ___ Nose (1-25) ___
Taste (1-25) ___ Finish (1-25) ___
Complexity (1-10) ___ Overall Impression (1-10) ___

SCORE ___

Parker's Heritage Collection Double Barreled Blend

Proof: 132.2
Age: 13 years and 15 years
Type: Straight
Style: Barrel Strength
Mash Bill: 78% corn, 10% rye, 12% malted barley
Color: Amber
Price: $$$$

Mike

Nose: Caramel, apples, allspice, and oak.
Taste: Caramel, apples, allspice, white pepper, and oak.
Finish: Long and dry with oak and pepper.

Susan

Nose: Warm cinnamon, chocolate, and caramel apples.
Taste: Cinnamon apple strudel with some milk chocolate and a touch of honey.
Finish: Long, warm, spicy.

Notes

Fall 2022 release.

My Score

Notes: ______________________________

____

____

Appearance (1-5) ____ Nose (1-25) ____
Taste (1-25) ____ Finish (1-25) ____
Complexity (1-10) ____ Overall Impression (1-10) ____

SCORE ____

Parker's Heritage Collection Promise of Hope

Proof: 96
Age: NAS
Type: Straight Bourbon
Style: Traditional
Mash Bill: Proprietary
Color: Amber
Price: $$$$

Mike

Nose: Very rich with caramel, dates, honey, sweet spices – nutmeg, cinnamon – and oak.

Taste: Almost like cognac, with pitted fruit – cherries and apricots. Cinnamon, fine tobacco and oak.

Finish: Long. Starts sweet with fruit and then gets a little more oak. Never looses sweetness.

Susan

Nose: Candies fruit and nuts. Sweet pecans, luscious vanilla and caramel. Sweet spices and new leather.

Taste: Chewy. Reminds me of cognac. Spicy tobacco and leather with rich vanilla and lots of peppery oak.

Finish: Long, warm and very, very smooth. Ends with sweet oak. Gorgeous.

Notes

Would be great with a steak and with a fine cigar afterwards. Gets better as it breathes! While there is no age statement on the bottle, the Heaven Hill website stated it was 10 years old.

My Score

Notes: ____

____

____

Appearance (1-5) ____ Nose (1-25) ____
Taste (1-25) ____ Finish (1-25) ____
Complexity (1-10) ____ Overall Impression (1-10) ____

SCORE ____

Square 6

Proof: 95
Age: NAS
Type: Straight
Style: Pot Distilled
Mash Bill: 52% corn, 35% rye, 13% malted barley
Color: Straw
Price: $$$

Mike

Nose: Lots of corn bread and vanilla with hints of fruit and oak.
Taste: Corn, vanilla, cardamom, nutmeg, and oak.
Finish: Long with lots of spice and oak.

Susan

Nose: Corn meal and fresh cut corn, a little brown sugar, nuts, and oak.
Taste: Candy corn with ripe apples sprinkled with brown sugar and nutmeg.
Finish: Long and warm with lingering sweetness.

Notes

Only available at the Evan Williams Bourbon Experience on Main Street in Louisville where it is distilled on site.

My Score

Notes: ______________________________

____

____

Appearance (1-5) ____ Nose (1-25) ____
Taste (1-25) ____ Finish (1-25) ____
Complexity (1-10) ____ Overall Impression (1-10) ____

SCORE ____

Square 6 Wheated

Proof: 105
Age: NAS
Type: Straight
Style: Wheated, Blend
Mash Bill: See notes.
Color: Straw
Price: $$$

Mike

Nose: Vanilla, bread dough, apples, nutmeg, oak.
Taste: Vanilla, apples, cinnamon, nutmeg, and oak.
Finish: Long with oak, spice, and lingering vanilla.

Susan

Nose: Wheat thins and corn bread with some dark fruit.
Taste: Plums and cherries, vanilla, and cinnamon.
Finish: Cinnamon all the way down.

Notes

Distilled at the Evan Williams Bourbon Experience on Whiskey Row (West Main Street) in Louisville. Blend of two mash bills – 74% corn, 16% wheat, 10% malted barley and 68% corn, 20% wheat, and 12% malted barley.

My Score

Notes: ______________________________

____

____

Appearance (1-5) ____ Nose (1-25) ____
Taste (1-25) ____ Finish (1-25) ____
Complexity (1-10) ____ Overall Impression (1-10) ____
SCORE ____

Very Special Old Fitzgerald

Proof: 90
Age: 12 years
Type: Straight Bourbon
Style: Wheated, Extra Aged
Mash Bill: 68% corn, 20% wheat, 10% malted barley
Color: Straw
Price: $$$

Mike

Nose: Great nose. Caramel, apples, pecans. Candied apple in a glass! Hints of milk chocolate and honey.

Taste: Nice warmth in the throat with caramel apples and peppery spice. Honey on the tongue quickly leads into other flavors.

Finish: Long and sweet with only a hint of oak tannins.

Susan

Nose: Honey, caramel, ripe apples, with rich vanilla aroma.

Taste: Caramel-coated fruit and nut bar. Honey on the tongue. Almost heartbreakingly smooth.

Finish: Long, honeyed with buttered sweet corn at the very end.

Notes

No longer made. But many large bourbon bars and vintage whiskey shops may still have some. Seek it out!

My Score

Notes: ______________________________

____

____

Appearance (1-5) ____ Nose (1-25) ____
Taste (1-25) ____ Finish (1-25) ____
Complexity (1-10) ____ Overall Impression (1-10) ____

SCORE ____

William Heavenhill Bottled-in-Bond

Proof: 100
Age: 132 months (11 years)
Type: Straight
Style: Bonded, Small Batch
Mash Bill: 78% corn, 10% rye, 12% malted barley
Color: Dark Straw
Price: $$$$

Mike

Nose: Caramel and peaches with oak tones.
Taste: Caramel corn and a hint of peach. Lots of oak, in a good way.
Finish: Dry and oaky with a hint of sweet caramel.

Susan

Nose: Caramel, dates, figs, light spice.
Taste: Dried fruit and oak. Think Fig Newton.
Finish: Long and warm with caramel at the end.

Notes

One of the limited editions sold in the Heaven Hill gift shops in Bardstown and Louisville. Each small batch was mingled from fewer than 12 barrels.

My Score

Notes: ______________________________

____

____

Appearance (1-5) ____ Nose (1-25) ____
Taste (1-25) ____ Finish (1-25) ____
Complexity (1-10) ____ Overall Impression (1-10) ____
SCORE ____

William Heavenhill Cask Strength

Proof:	135.6
Age:	15 years
Type:	Straight
Style:	Extra Aged, Barrel Proof
Mash Bill:	75% corn, 13% rye, 12% malted barley
Color:	Rich Amber
Price:	$$$$

Mike

Nose: Chocolate caramel chew. Some apricot and spice. Water releases spice.

Taste: Chocolate and caramel with apricots and berries with oak. Water brings out the chocolate.

Finish: Dry with oak and chocolate. Water makes finish less oaky, more chocolate.

Susan

Nose: Caramel, vanilla, chocolate, dark dried fruit, pecans.

Taste: Vanilla, saddle leather, milk chocolate, pitted fruit. Water amplifies fruit and nuts.

Finish: Long, warm, sweet and complex.

Notes

One of the limited editions sold in the Heaven Hill gift shops in Bardstown and Louisville. Retail about $250. You may find it in the top bourbon bars in big cities. Know there were only about 350 bottles.

My Score

Notes: ____

____

____

Appearance (1-5)	____	Nose (1-25)	____
Taste (1-25)	____	Finish (1-25)	____
Complexity (1-10)	____	Overall Impression (1-10)	____

SCORE ____

Heaven Hill Additional Releases

Name: ______________________________

Distillery: ______________________________

Proof: __________ **Mash Bill:** __________
Age: __________ __________
Type: __________ **Color:** __________
Style: __________ **Price:** __________

My Score

Notes: ______________________________

____

____

Appearance (1-5) ____ Nose (1-25) ____
Taste (1-25) ____ Finish (1-25) ____
Complexity (1-10) ____ Overall Impression (1-10) ____
SCORE ____

Name: ______________________________

Distillery: ______________________________

Proof: __________ **Mash Bill:** __________
Age: __________ __________
Type: __________ **Color:** __________
Style: __________ **Price:** __________

My Score

Notes: ______________________________

____

____

Appearance (1-5) ____ Nose (1-25) ____
Taste (1-25) ____ Finish (1-25) ____
Complexity (1-10) ____ Overall Impression (1-10) ____
SCORE ____

Lexington Brewing & Distilling

https://lexingtonbrewingco.com

Bourbon Distilleries:

- Town Branch Distillery, Lexington, Ky.

Town Branch opened in 1999. It was owned by Pearce Lyons' Alltech, an animal feed supplement company. Lyons was also a distiller, having learned the science in his native Ireland, which is why Town Branch has pot stills and also why, in addition to bourbon, it makes two expressions of malt whiskey. As reflected by the company name, this is also a brewery and not surprisingly, bourbon ex-bourbon barrels are used to age its Kentucky Bourbon Barrel Ale and some other beers.

Town Branch

Proof: 80
Age: NAS
Type: Straight
Style: Craft
Mash Bill: 72% corn, 15% rye, 13% malted barley
Color: Pale Straw
Price: $$

Mike

Nose: Corn and vanilla with a hint of spice. Very simple nose.
Taste: Candy corn and pepper spice with a bit of oak. A bit thin in the mouth.
Finish: Starts peppery and fades quickly.

Susan

Nose: Simple. Corn, light caramel.
Taste: Corn predominates, followed by caramel, a little vanilla and faint nuttiness.
Finish: Short and slightly bitter with tannins at the end.

Notes

An all-around rather light-bodied whiskey for a bourbon. Produced at Town Branch Distillery in Lexington, Kentucky. It is owned by All-Tech.

My Score

Notes: ____

____

____

Appearance (1-5) ____ Nose (1-25) ____
Taste (1-25) ____ Finish (1-25) ____
Complexity (1-10) ____ Overall Impression (1-10) ____
SCORE ____

Town Branch Sherry Cask Finished

Proof:	100
Age:	NAS*
Type:	Straight
Style:	Finished
Mash Bill:	72% corn, 13% rye, 15% malted barley
Color:	Dark Straw
Price:	$$

Mike

Nose: Caramel, dates, light baking spice, and oak.

Taste: Caramel, cherries, dates, baking spices, notes of chocolate and oak.

Finish: Long with oak, spice, and lingering fruit notes.

Susan

Nose: Apples, cornbread, a hint of mint, and some new leather.

Taste: Sweet, almost grapey. Then a dash of baking spice over a caramel base.

Finish: Dries to some light oak and very light pepper.

Notes

Lexington Brewing & Distilling Co., Kentucky. Aged 8 years before 1 additional year in the Oloroso sherry cask.

My Score

Notes: ______________________________

____

____

Appearance (1-5) ____ Nose (1-25) ____

Taste (1-25) ____ Finish (1-25) ____

Complexity (1-10) ____ Overall Impression (1-10) ____

SCORE ____

Town Branch Single Barrel Reserve

Proof:	109.2
Age:	6 years
Type:	Straight
Style:	Cask Strength
Mash Bill:	72% corn, 13% rye, 15% malted barley
Color:	Straw
Price:	$$

Mike

Nose: Vanilla and malty grain notes, apples and a hint of oak.

Taste: Vanilla, malted milk balls with a hint of chocolate, black pepper, and oak.

Finish: Medium long with oak and pepper.

Susan

Nose: Herbal, malty, with a little fruit and a bit of vanilla.

Taste: More vanilla with some berries and new leather.

Finish: Long and warm.

Notes

Water brings out a little apple and cinnamon.

My Score

Notes: ____

____

____

Appearance (1-5) ____ Nose (1-25) ____

Taste (1-25) ____ Finish (1-25) ____

Complexity (1-10) ____ Overall Impression (1-10) ____

SCORE ____

Town Branch True Cask

Proof:	108.1
Age:	NAS
Type:	Straight
Style:	Blend
Mash Bill:	72% corn, 13% rye, 15% malted barley + another not available
Color:	Dark Straw
Price:	$$

Mike

Nose: Caramel, apples, hint of spice, and oak.
Taste: Caramel, green apples, hints of spice and oak.
Finish: Long and dry with oak and pepper.

Susan

Nose: Very nutty with a little honey in the background.
Taste: Medley of nuts – cashews, walnuts, hazelnuts – with some caramel and a touch of baking spice.
Finish: Long and oaky.

Notes

Blend of two Kentucky mash bills.

My Score

Notes: ______________________________

____

____

Appearance (1-5) ____ Nose (1-25) ____
Taste (1-25) ____ Finish (1-25) ____
Complexity (1-10) ____ Overall Impression (1-10) ____

SCORE ____

Lexington Brewing & Distilling Additional Releases

Name: ______________________________

Distillery: ______________________________

Proof: ____________ **Mash Bill:** ____________
Age: ____________ ____________
Type: ____________ **Color:** ____________
Style: ____________ **Price:** ____________

My Score

Notes: ______________________________

____

____

Appearance (1-5) ____ Nose (1-25) ____
Taste (1-25) ____ Finish (1-25) ____
Complexity (1-10) ____ Overall Impression (1-10) ____
SCORE ____

Name: ______________________________

Distillery: ______________________________

Proof: ____________ **Mash Bill:** ____________
Age: ____________ ____________
Type: ____________ **Color:** ____________
Style: ____________ **Price:** ____________

My Score

Notes: ______________________________

____

____

Appearance (1-5) ____ Nose (1-25) ____
Taste (1-25) ____ Finish (1-25) ____
Complexity (1-10) ____ Overall Impression (1-10) ____
SCORE ____

Lux Row

www.luxrowdistillers.com

Bourbon Distilleries:

- Lux Row, Bardstown, Ky.

Founded in St. Louis in1958. in Luxco distributed many brands that it had contract distilled in Kentucky, most notably by Heaven Hill. Sixty years on, the company built its own Kentucky facility, which opened in 2018. Both the St. Louis Company and Lux Row Distillery were acquired in 2021 by MPG, the distilling giant in Lawrenceburg, Indiana from which many independent bottlers source their whiskey.

Blood Oath

Proof: 98.6
Age: NAS
Type: Straight
Style: Small Batch
Mash Bill: 68% corn, 20% wheat, 12% malted barley
Color: Straw
Price: $$$$

Mike

Nose: Caramel and leather with a hint of honeysuckle.
Taste: Caramel and apricots with some pepper spice.
Finish: Woody and a bit astringent.

Susan

Nose: Banana, apricot, vanilla, sweet spice.
Taste: Light caramel, fruit and sweet spices.
Finish: Warm with dash of black pepper.

Notes

Very smooth sip, especially give the proof, which is a play on blood temperature. This was from "Pact No. 1" (a mingling of three different mash bills) and batch number 1. Bottled 2-20-15. A Pact No. 2 was released in 2016 featuring a different combination of mash bills.

My Score

Notes: ______________________________

____

____

Appearance (1-5) ____ Nose (1-25) ____
Taste (1-25) ____ Finish (1-25) ____
Complexity (1-10) ____ Overall Impression (1-10) ____
SCORE ____

David Nicholson 1843

Proof: 100
Age: NAS
Type: Straight
Style: Traditional
Mash Bill: 68% corn, 20% wheat, 12% malted barley
Color: Light Straw
Price: $$

Mike

Nose: Corn and vanilla with a hint of banana.
Taste: Corn husk and vanilla. Thin for 100 proof.
Finish: Short and dry. Corn husk lingers.

Susan

Nose: Toffee and orange zest.
Taste: Some caramel and corn with black pepper.
Finish: Herbal, peppery.

Notes

Brand originated with Stitzel-Weller. Sourced from Kentucky. Distributed by Luxco of St. Louis, Missouri.

My Score

Notes: ______________________________________

____

____

Appearance (1-5) ____ Nose (1-25) ____
Taste (1-25) ____ Finish (1-25) ____
Complexity (1-10) ____ Overall Impression (1-10) ____
SCORE ____

Ezra B

Proof: 99
Age: 12 years
Type: Straight
Style: Single Barrel, Extra Aged
Mash Bill: 68% corn, 20% wheat, 12% malted barley
Color: Light Amber
Price: $$$

Mike

Nose: Old leather and caramel. Apricot with pepper spice. Caramel gets stronger as it breathes.
Taste: Nice caramel and fruit – apricot, apple – apples with pepper spice and fine leather and tobacco.
Finish: Long and dry with oak and pepper. A hint of fruit sweetness.

Susan

Nose: Caramel corn, vanilla, sweet oak, and some dried stone fruit.
Taste: Caramel, cherries, ripe apple, vanilla. Whiff of tobacco leaf.
Finish: Long and warm with pleasant amounts of spice and fruit.

Notes

No longer made but can be found in bourbon bars. Made by Heaven Hill and bottled by Luxco, St. Louis, Missouri.

My Score

Notes: ______________________________________

____

____

Appearance (1-5) ____ Nose (1-25) ____
Taste (1-25) ____ Finish (1-25) ____
Complexity (1-10) ____ Overall Impression (1-10) ____

SCORE ____

Old Ezra

Proof: 101
Age: 7 years
Type: Straight
Style: Traditional
Mash Bill: 68% corn, 20% wheat, 12% malted barley
Color: Dark Straw
Price: $$

Mike

Nose: Caramel and vanilla with tobacco and leather and sweet spices – nutmeg?
Taste: Nice mouthfeel. Almost chewy with caramel and peppery spices. Apple fruit.
Finish: Medium long with sweet fruit at first drying to oak tannins.

Susan

Nose: Fruity. Light caramel and roasted corn and pie spices.
Taste: Caramel corn, pitted fruit with a bit cocoa and pepper.
Finish: Long and warm and chewy without being hot. Ends in oak.

Notes

Not especially complex, but a satisfying sip. Produced by Heaven Hill in Louisville, Kentucky. Bottled and distributed by Luxco of St. Louis, Missouri.

My Score

Notes: ______________________________
____
____

Appearance (1-5) ____ Nose (1-25) ____
Taste (1-25) ____ Finish (1-25) ____
Complexity (1-10) ____ Overall Impression (1-10) ____

SCORE ____

Old Ezra Barrel Strength

Proof: 117
Age: 7 years
Type: Straight
Style: Cask Strength
Mash Bill: 78% corn, 12% rye, 10% malted barley
Color: Light Amber
Price: $$$

Mike

Nose: Vanilla and berries with some oak and leather.
Taste: Vanilla, blackberries, pepper spice, and oak.
Finish: Long and spicy with pepper and oak.

Susan

Nose: Sweet corn, a little milk chocolate, and vanilla icing.
Taste: Cherries and vanilla with some nutmeg and more milk chocolate.
Finish: Long and complex.

Notes

Very smooth for the proof. Water releases more fruit and chocolate.

My Score

Notes: ______________________________

____

____

Appearance (1-5) ____ Nose (1-25) ____
Taste (1-25) ____ Finish (1-25) ____
Complexity (1-10) ____ Overall Impression (1-10) ____

SCORE ____

Rebel Reserve

Proof:	90.6
Age:	NAS
Type:	Straight
Style:	Wheated
Mash Bill:	Not available
Color:	Pale Straw
Price:	$$$

Mike

Nose: Simple. Caramel and vanilla with bit of oak. Very light cotton candy in the background.

Taste: Caramel corn and ripe apples with a bit of hazelnuts or pecans.

Finish: Long and dry with oak tannins and pepper spice.

Susan

Nose: Very light. Vanilla and faint corn. Sugar is lurking underneath.

Taste: Caramel candy corn, sweet cereal. Faint nuttiness and dash of pepper.

Finish: Fairly long, drying to pepper spice.

Notes

Moderately-priced bottle will be interesting to compare to other wheated bourbons.

My Score

Notes: ____

____

____

Appearance (1-5) ____ Nose (1-25) ____

Taste (1-25) ____ Finish (1-25) ____

Complexity (1-10) ____ Overall Impression (1-10) ____

SCORE ____

Rebel Yell

Proof: 80
Age: NAS
Type: Straight
Style: Wheated
Mash Bill: Not available
Color: Pale Straw
Price: $

Mike

Nose: Very simple. Vanilla and corn with a hint of oak.
Taste: Thin and watery. Vanilla and corn with some hazelnut.
Finish: A bit dry with oak tannins. Lasts longer than expected.

Susan

Nose: Very light, with notes of vanilla, candy corn, perhaps a whiff of mint.
Taste: Light nuttiness and faint spice support the vanilla.
Finish: Medium long with pepper spice.

Notes

Once upon a time, this was a 90 proof bourbon. Distilled by Heaven Hill in Louisville, Kentucky. Distributed by Luxco Brands of St. Louis, Missouri.

My Score

Notes: ____________________________________

____

____

Appearance (1-5) ____ Nose (1-25) ____
Taste (1-25) ____ Finish (1-25) ____
Complexity (1-10) ____ Overall Impression (1-10) ____

SCORE ____

Lux Row Additional Releases

Name: ______________________

Distillery: ______________________

Proof: __________ **Mash Bill:** __________
Age: __________ __________
Type: __________ **Color:** __________
Style: __________ **Price:** __________

My Score

Notes: ______________________

____

____

Appearance (1-5) ___ Nose (1-25) ___
Taste (1-25) ___ Finish (1-25) ___
Complexity (1-10) ___ Overall Impression (1-10) ___
SCORE ___

Name: ______________________

Distillery: ______________________

Proof: __________ **Mash Bill:** __________
Age: __________ __________
Type: __________ **Color:** __________
Style: __________ **Price:** __________

My Score

Notes: ______________________

____

____

Appearance (1-5) ___ Nose (1-25) ___
Taste (1-25) ___ Finish (1-25) ___
Complexity (1-10) ___ Overall Impression (1-10) ___
SCORE ___

Michter's

www.michters.com

Bourbon Distilleries:

- Michter's Distillery, Shively, Ky.
- Michter's Fort Nelson Distillery, Louisville, Ky.

Chatham Imports rescued the Michter's brand from the dustpan of history in the 1990s. The brand was created in Pennsylvania in the 1950s and survived until the 1980s when the distillery went dark and the owners walked away from it, leaving the distillery and some inventory to the bank. The present owners saved the brand, found old barrels for sale and contract distilled whiskeys. Since 2015, Michter's has been distilling its own bourbon and rye at its new distillery in Shively, Kentucky which happens to be right next door to Brown-Forman's distillery.

It also operates a micro-distillery, Fort Nelson, on Louisville's Whiskey Row using the pot still from the old Pennsylvania distillery.

Bomberger's Declaration Kentucky Straight Bourbon

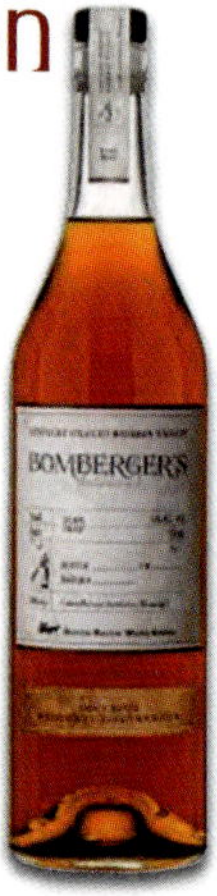

Proof: 108
Age: NAS
Type: Straight
Style: Small Batch
Mash Bill: Not available
Color: Dark Straw
Price: $$$$

Mike

Nose: Caramel, pears, baking spices, oak.
Taste: Caramel, pears, cinnamon, nutmeg, allspice, oak.
Finish: Medium long with oak, spice, and lingering caramel.

Susan

Nose: Caramel, cashews, ripe apples.
Taste: Sweet with dark brown sugar, apples, roasted cashews, and cinnamon sugar.
Finish: Warm, smooth, very balanced.

Notes

These annual releases are pricey, but so not hesitate to buy when you see it!

My Score

Notes: ____________________________________

____

____

Appearance (1-5) ____ Nose (1-25) ____
Taste (1-25) ____ Finish (1-25) ____
Complexity (1-10) ____ Overall Impression (1-10) ____
SCORE ____

Michter's US*1 Barrel Strength

Proof: 110.6
Age: NAS
Type: Straight
Style: Barrel Strength
Mash Bill: Not available
Color: Amber
Price: $$$$

Mike

Nose: Caramel apple or ripe pear with a little oak and spices.
Taste: Caramel and ripe apple, stone fruit and nutmeg.
Finish: Long and dry with oak and spice.

Susan

Nose: Toffee, vanilla, orange blossom. Whiff of cinnamon.
Taste: Sweet corn and lashings of caramel and vanilla with dried fruit.
Finish: Impressively smooth at this proof. Dries to sweet oak.

Notes

Water brings out fruit, cinnamon, and nutmeg and dries the finish. Excellent "book bourbon." Savor a sip after each chapter you read of your book of the evening. Released only in Kentucky.

My Score

Notes: ______________________________

____

____

Appearance (1-5) ____ Nose (1-25) ____
Taste (1-25) ____ Finish (1-25) ____
Complexity (1-10) ____ Overall Impression (1-10) ____

SCORE ____

Michter's US*1 Small Batch Bourbon

Proof: 91.4
Age: NAS
Type: Not Straight
Style: Small Batch
Mash Bill: Not available
Color: Light Straw
Price: $$

Mike

Nose: Caramel, toffee and fine leather. A bit of fruit – ripe pears.
Taste: Sweet caramel and apricots with pepper spice. Not overly complex. Nice mouthfeel with no burn.
Finish: Long. Starts sweet but gets dry with oak tannins and pepper spice.

Susan

Nose: Toffee, lots of pepper spice and vanilla with touch of ripe apple and honey.
Taste: Carmel corn and pepper spice with a tiny bit of green apple. Buttery mouthfeel.
Finish: Long and dry. Very peppery.

Notes

Try it in a Manhattan with a very good quality vermouth. Experiment with bitters to bring out the fruit notes. Michter's bourbons are currently sourced from other distilleries.

My Score

Notes: ______________________________

____

____

Appearance (1-5) ____ Nose (1-25) ____
Taste (1-25) ____ Finish (1-25) ____
Complexity (1-10) ____ Overall Impression (1-10) ____
SCORE ____

Michter's US*1 Small Batch Toasted Barrel Finished

Proof: 91.4
Age: NAS
Type: Not Straight
Style: Small Batch #14I539
Mash Bill: Not available
Color: Dark Straw
Price: $$$

Mike

Nose: Butterscotch and caramel. Sweet corn and butter. Fine leather and oak.
Taste: Butterscotch, vanilla, corn and oak with some sweet honey and apple.
Finish: Long. Starts sweet and the oak kicks in to dry it out. Very good.

Susan

Nose: Wood smoke, vanilla, leather, and peaches.
Taste: Vanilla ice cream with fruit and Small cocoa note and hint of mint. It's a sundae!
Finish: Long and sweet, with a fine smooth oak note a the end.

Notes

Great with a steak and a pipe afterwards.

My Score

Notes: ____
____
____

Appearance (1-5) ____ Nose (1-25) ____
Taste (1-25) ____ Finish (1-25) ____
Complexity (1-10) ____ Overall Impression (1-10) ____
SCORE ____

Michter's 10-Year-Old

Proof:	94.4
Age:	10
Type:	Straight
Style:	Small Batch, Extra Aged
Mash Bill:	Not available
Color:	Very Dark Straw
Price:	$$$

Mike

Nose: Caramel and milk chocolate with some apricot fruit. Just hint of spice.

Taste: Apricot and caramel with some sweet spice. Nutmeg and cinnamon with fine tobacco.

Finish: Long and dry with oak tannins and pepper. tannins and pepper spice.

Susan

Nose: Honeycomb, vanilla, touch of ripe cherries.

Taste: Honey, pitted fruit, sweet corn. Very mellow.

Finish: Long and dry with just a little pepper at the very end.

Notes

Addition of a little water makes the finish much sweeter. The fruit carries on longer.

My Score

Notes: ______________________________

____

____

Appearance (1-5)	____	Nose (1-25)	____
Taste (1-25)	____	Finish (1-25)	____
Complexity (1-10)	____	Overall Impression (1-10)	____

SCORE ____

Michter's 20-Year-Old

Proof: 114.2
Age: NAS
Type: Straight
Style: Single Barrel, #37/38, Extra Aged
Mash Bill: Not available
Color: Amber
Price: $$$$

Mike

Nose: Crème brûlée, caramel and vanilla with some oak and sweet spice.
Taste: Caramel and apricot. Dates with oak wood and pepper spice.
Finish: Long and dry with oak tannins and pepper spice.

Susan

Nose: Vanilla, caramel, heavy honey with underlying spice and oak.
Taste: Serious sweet oak from the aging. But mellow with pitted fruit and honeycomb.
Finish: Long and dry, but very, very smooth.

Notes

Very smooth and elegant sip for its age and proof. A splash of water amplifies the sweet notes.

My Score

Notes: ____________________________________

____

____

Appearance (1-5) ____ Nose (1-25) ____
Taste (1-25) ____ Finish (1-25) ____
Complexity (1-10) ____ Overall Impression (1-10) ____
SCORE ____

Michter's Additional Releases

Name: ______________________________

Distillery: ______________________________

Proof: ____________ **Mash Bill:** ____________
Age: ____________ ____________
Type: ____________ **Color:** ____________
Style: ____________ **Price:** ____________

My Score

Notes: ______________________________

____

____

Appearance (1-5) ____ Nose (1-25) ____
Taste (1-25) ____ Finish (1-25) ____
Complexity (1-10) ____ Overall Impression (1-10) ____
SCORE ____

Name: ______________________________

Distillery: ______________________________

Proof: ____________ **Mash Bill:** ____________
Age: ____________ ____________
Type: ____________ **Color:** ____________
Style: ____________ **Price:** ____________

My Score

Notes: ______________________________

____

____

Appearance (1-5) ____ Nose (1-25) ____
Taste (1-25) ____ Finish (1-25) ____
Complexity (1-10) ____ Overall Impression (1-10) ____
SCORE ____

Sazerac Company

Bourbon Distilleries:

- A. Smith Bowman Distillery, Fredericksburg, Va.
- Barton 1792 Distillery, Bardstown, Ky.
- Buffalo Trace Distillery, Frankfort, Ky.

Thomas Handy purchased the Sazerac Coffee House in 1869 and eventually became the Sazerac Company. They purchased the Ancient Age distillery in 1992 and changed the name to Buffalo Trace in 1997. In 2003 Sazerac Company purchased the A. Smith Bowman Distillery and in 2009 they purchased Barton Brands and the Barton 1792 distillery.

Buffalo Trace started life as the OFC distillery founded by E. H. Taylor, Jr. in 1870. Taylor loses control of the distillery to the firm of Gregory and Stagg in 1879 and the distillery eventually becomes known as the Geo. T. Stagg distillery. During prohibition the distillery was tied to Schenley distilleries Corporation who purchase the distillery outright in 1933. It remains part of Schenley until 1984 when it becomes an independent distillery.

Tom Moore founded Barton 1792 in 1889. The distillery was closed during Prohibition but the Moore family re-opened it after Repeal. In 1944 Oscar Getz purchased the distillery and changed the name to the Barton Distillery. Getz retires in 1982 and the distillery was sold in 1983. It changed hands several times before it finally landed with Sazerac in 2009.

A. Smith Bowman

www.asmithbowman.com

The A. Smith Bowman distillery was found in 1934 in Fairfax County, Va. In 1988 the distillery was relocated to its present site and retained only the pot still doubler, purchasing new make from other distilleries to double distill for their Bourbon brands. In 2003 the distillery was sold to Sazerac.

Bowman Brothers

Proof: 90
Age: NAS
Type: Straight
Style: Small Batch
Mash Bill: Proprietary
Color: Dark Straw
Price: $$$

Mike

Nose: Butter, caramel, and corn with a little pepper spice.
Taste: Candy corn and pepper spice with a bit of oak.
Finish: Long and dry with oak tannins and pepper spice.

Susan

Nose: Vanilla, honey, pleasant apple fruit and a tickling of pepper.
Taste: Surprisingly spicy after the sweet nose. Sweet corn and vanilla, too.
Finish: Sweet honey fades to pepper spice.

Notes

Comes from A. Smith Bowman Distillery, Fredericksburg, Virginia.

My Score

Notes: ________________________________

____

____

Appearance (1-5) ____ Nose (1-25) ____
Taste (1-25) ____ Finish (1-25) ____
Complexity (1-10) ____ Overall Impression (1-10) ____
SCORE ____

John J. Bowman

Proof: 100
Age: NAS
Type: Straight
Style: Single Barrel
Mash Bill: Proprietary
Color: Dark Straw
Price: $$$

Mike

Nose: Honey and vanilla with corn and oak.
Taste: Honey and corn, Candy corn with some oak and sweet spices.
Finish: Long and dry with spicy pepper and oak tannins.

Susan

Nose: Vanilla and apples, hint of sweet spice.
Taste: Buttery, yeasty, and toasty. Beautifully balanced.
Finish: Long and sweet. barely dries.

Notes

Comes from A. Smith Bowman Distillery, Fredericksburg, Virginia. Very smooth, elegant bourbon.

My Score

Notes: ______________________________

____

____

Appearance (1-5) ____ Nose (1-25) ____
Taste (1-25) ____ Finish (1-25) ____
Complexity (1-10) ____ Overall Impression (1-10) ____

SCORE ____

A. Smith Bowman Additional Releases

Name: ______________________________

Distillery: ______________________________

Proof: ____________ **Mash Bill:** ____________
Age: ____________ ____________
Type: ____________ **Color:** ____________
Style: ____________ **Price:** ____________

My Score

Notes: ______________________________

____

____

Appearance (1-5) ____ Nose (1-25) ____
Taste (1-25) ____ Finish (1-25) ____
Complexity (1-10) ____ Overall Impression (1-10) ____
SCORE ____

Name: ______________________________

Distillery: ______________________________

Proof: ____________ **Mash Bill:** ____________
Age: ____________ ____________
Type: ____________ **Color:** ____________
Style: ____________ **Price:** ____________

My Score

Notes: ______________________________

____

____

Appearance (1-5) ____ Nose (1-25) ____
Taste (1-25) ____ Finish (1-25) ____
Complexity (1-10) ____ Overall Impression (1-10) ____
SCORE ____

Barton 1792

www.1792distillery.com

Uses three different mash bills — BW 4, BW 6, BW 8 — including for bourbons contract distilled for independent bottlers. But no information released about grain proportions. That said Modern Thirst has published 75% corn, 15% rye, and 10 malted barley as the mash bill for the distillery's labels. Again, use this information advisedly.

1792 Full Proof

Proof: 125
Age: ~ 8 ½ years
Type: Straight
Style: Non-Chill Filtered
Mash Bill: Not available
Color: Dark Straw
Price: $$$

Mike

Nose: Brown sugar and fruit.
Taste: Apples and pears. Cinnamon and cloves.
Finish: Dry oak and sweet fruit.

Susan

Nose: A caramel bomb with chocolate sprinkles.
Taste: Caramel, chocolate, fruit, and cloves.
Finish: Long, sweet, warm.

Notes

When the bourbon came out of the barrel it was 132-142 proof. Distilled water was used to return the bourbon to the proof at which it entered the barrel. Hence, "full proof."

My Score

Notes: ____________________________________

____

____

Appearance (1-5) ____ Nose (1-25) ____
Taste (1-25) ____ Finish (1-25) ____
Complexity (1-10) ____ Overall Impression (1-10) ____
SCORE ____

1792 High Rye

Proof: 94.3
Age: NAS
Type: Straight
Style: Small Batch
Mash Bill: Higher rye than usual expression
Color: Dark Straw
Price: $$$

Mike

Nose: Candied fruit – cherries and dates.
Taste: Tobacco, cherries and dates, caramel, and sweet spice.
Finish: Long, dry, and spicy.

Susan

Nose: Caramel, spice, candy corn, sweet mint.
Taste: Surprisingly sweet for high rye content. Bing cherries and sweet mint.
Finish: Long with sweet oak and a whiff of honey.

Notes

The definition of "small batch" here is 80-120 barrels. A remarkably mellow sip for a robust rye mash bill.

My Score

Notes: ____

____

____

Appearance (1-5) ____ Nose (1-25) ____
Taste (1-25) ____ Finish (1-25) ____
Complexity (1-10) ____ Overall Impression (1-10) ____

SCORE ____

1792 Port Finish

Proof: 88.9
Age: 8 years
Type: Straight
Style: Special Finish
Mash Bill: Not available
Color: Light Amber
Price: $$$

Mike

Nose: Fruit and caramel. More port than bourbon.
Taste: Fruity with raisins and berries, some caramel and a hint of sweet spices.
Finish: Short and fruity with a hint of oak.

Susan

Nose: Port sweetness evident with some vanilla and honey.
Taste: Berries, caramel, sweet spices. Touch of licorice.
Finish: Long and fruity with sweet oak.

Notes

The bourbon spent six years in oak barrels and two more in port barrels. Water brings out the fruit.

My Score

Notes: ____

____

____

Appearance (1-5) ____ Nose (1-25) ____
Taste (1-25) ____ Finish (1-25) ____
Complexity (1-10) ____ Overall Impression (1-10) ____

SCORE ____

1792 Small Batch

Proof: 93.7
Age: 8 years
Type: Straight Bourbon
Style: Small Batch
Mash Bill: BW 8
Color: Amber
Price: $$

Mike

Nose: Very complex. Vanilla/caramel, fine leather, sweet spice – nutmeg. Very ripe apple.

Taste: Nice mouthfeel. Apples and caramel with sweet spices that become a bit peppery. A little oak just before the finish.

Finish: Long and spicy with peppery spice and caramel. Very nice.

Susan

Nose: Vanilla, black pepper, new leather, cinnamon and light oak. Multilayered.

Taste: Vanilla and pepper. More sweet spices. Faint flavor of dark fruit mid-palate, but more spicy than fruity.

Finish: Long finish starts sweet and ends with black Pepper.

Notes

A good bourbon to accompany food. Enjoy with a variety of entrees. Fine with a fruity dessert. Good neat or mixed in cocktails, too.

My Score

Notes: ______________________________________

____

____

Appearance (1-5) ____ Nose (1-25) ____
Taste (1-25) ____ Finish (1-25) ____
Complexity (1-10) ____ Overall Impression (1-10) ____

SCORE ____

1792 Sweet Wheat

Proof: 91.2
Age: NAS
Type: Straight
Style: Wheated
Mash Bill: Not available
Color: Straw
Price: $$$

Mike

Nose: Caramel and apples with a little oak and pecans.
Taste: Caramel apple fruit with some vanilla and hazelnuts.
Finish: Starts sweet with some caramel, but slowly gets some oak.

Susan

Nose: Sweet corn, ripe apples, nutmeg.
Taste: Lots of vanilla, some pecans, sweet spices.
Finish: Smooth and drying to sweet oak.

Notes

Lovers of wheated bourbons will enjoy this. Limited release in 2015, but may still be on bourbon bar shelves.

My Score

Notes: ______________________________

____

____

Appearance (1-5) ____ Nose (1-25) ____
Taste (1-25) ____ Finish (1-25) ____
Complexity (1-10) ____ Overall Impression (1-10) ____

SCORE ____

Colonel Lee

Proof: 80
Age: NAS
Type: Straight Bourbon
Style: Traditional
Mash Bill: BW 4
Color: Light Straw
Price: $

Mike

Nose: Vanilla and cherries/apricots, with a little leather and oak. Very light on the nose.
Taste: Sweet vanilla and corn with a bit of honey and pepper spice. A bit of oak and tobacco.
Finish: Starts sweet, but gets drier with the oak tannins and pepper.

Susan

Nose: Light caramel, faint nuttiness, oak notes.
Taste: Caramel corn, pecans, beeswax. Vanilla asserts itself in the middle of the mouth.
Finish: Medium finish with a peppery kick at the very end.

Notes

An inexpensive bourbon that is a natural to serve mixed with cola or ginger ale.

My Score

Notes: ____

____

____

Appearance (1-5) ____ Nose (1-25) ____
Taste (1-25) ____ Finish (1-25) ____
Complexity (1-10) ____ Overall Impression (1-10) ____

SCORE ____

Early Times Bottled-in-Bond

Proof: 100
Age: 4 years
Type: Straight
Style: Bonded
Mash Bill: 79% corn, 11% rye, 10% malted barley
Color: Straw
Price: $

Mike

Nose: Fritos corn chips, vanilla, pepper, oak.
Taste: Vanilla, Fritos, pepper, and oak.
Finish: Medium long with oak and pepper.

Susan

Nose: Crème brulee, ripe apple, light nutmeg.
Taste: Savory. Herbs, leather, caramel corn, and some pepper and oak.
Finish: Medium long with oak and pepper.

Notes

Brown-Forman sold the brand to Sazerac in 2020 and it is now based at Barton in Bardstown.

My Score

Notes: ______________________________

____

____

Appearance (1-5) ___ Nose (1-25) ___
Taste (1-25) ___ Finish (1-25) ___
Complexity (1-10) ___ Overall Impression (1-10) ___
SCORE ___

Kentucky Gentleman

Proof: 80
Age: NAS
Type: Bourbon a Blend
Style: Traditional
Mash Bill: BW 8
Color: Light Straw
Price: $

Mike

Nose: Vanilla and corn with some leather, corn and toasted marshmallow.
Taste: Very light. Corn and vanilla with a little pepper spice.
Finish: Short and dry. Very disappointing on the finish.

Susan

Nose: Very simple – alcohol and wood with an undertone of vanilla.
Taste: Faint wood smoke and little vanilla and corn.
Finish: Flavors flash by. Very short and warm.

Notes

Drink this with mixers such as ginger ale or Coke.

My Score

Notes: ____________________________________

____

____

Appearance (1-5) ____ Nose (1-25) ____
Taste (1-25) ____ Finish (1-25) ____
Complexity (1-10) ____ Overall Impression (1-10) ____

SCORE ____

Kentucky Tavern

Proof: 80
Age: NAS
Type: Straight Bourbon
Style: Traditional
Mash Bill: BW 4
Color: Very Light Straw
Price: $

Mike

Nose: Caramel corn and oak with a hint of tobacco.
Taste: Very light with sweet corn and vanilla. A bit or fruit – apricot or peach – and pepper spice.
Finish: Starts sweet, but gets drier as the oak tannins kick in with some sweet spice.

Susan

Nose: Vanilla, sweet corn, dash of cinnamon.
Taste: Stone fruitiness (cherries?) with sweet caramel, vanilla and sweet spices.
Finish: Medium length finish that dries to the end. Smooth throughout; no burn.

Notes

Nice sip for the price. Louisville chef Kathy Cary, the owner of famed restaurant Lilly's, uses Kentucky Tavern to make her savory KT steak sauce served with the beef tenderloin sandwich.

My Score

Notes: ____________________________________

____

____

Appearance (1-5) ____ Nose (1-25) ____
Taste (1-25) ____ Finish (1-25) ____
Complexity (1-10) ____ Overall Impression (1-10) ____

SCORE ____

Ten High

Proof: 80
Age: 4 years
Type: Bourbon a Blend
Style: Traditional
Mash Bill: BW 4
Color: Pale Straw
Price: $

Mike

Nose: Alcohol, light vanilla. Very simple nose.
Taste: Slightly musty. A little vanilla and corn.
Finish: Short and sweet. A little mustiness.

Susan

Nose: Very simple. Sweet corn and vanilla.
Taste: Sweet vanilla and corn. A little hot mid-palate.
Finish: Light, quick finish with touch of mustiness But no burn.

Notes

Because this is a blend, a high proportion is grain neutral spirits, which keeps the price bottom shelf. Inexpensive whiskey to use in your bourbon and Coke.

My Score

Notes: ____

____

____

Appearance (1-5) ____ Nose (1-25) ____
Taste (1-25) ____ Finish (1-25) ____
Complexity (1-10) ____ Overall Impression (1-10) ____

SCORE ____

Very Old Barton 80

Proof: 80
Age: NAS
Type: Straight Bourbon
Style: Traditional
Mash Bill: BW 4
Color: Amber
Price: $

Mike

Nose: Sweet caramel and fine leather. Apples and pears with some oak and tobacco.
Taste: A bit thin and watery. Sweet caramel and pears with oak tannins.
Finish: Very sweet at the start and gets peppery and dry. A bit short, but pleasant.

Susan

Nose: Caramel, apricot and sweet corn. A little spiced oak.
Taste: Oak up front followed by caramel and dried fruit.
Finish: Warm and smooth as it moves from sweet to spicy.

Notes

Good mixing bourbon for either cocktails or with carbonated sodas.

My Score

Notes: ____________________

____

____

Appearance (1-5) ____ Nose (1-25) ____
Taste (1-25) ____ Finish (1-25) ____
Complexity (1-10) ____ Overall Impression (1-10) ____

SCORE ____

Very Old Barton 86

Proof: 86
Age: 6 years
Type: Straight Bourbon
Style: Traditional
Mash Bill: BW 6
Color: Dark Straw
Price: $

Mike

Nose: Sweet caramel and leather. Corn and vanilla with a hint of oak. As it breathes, there is more caramel.
Taste: Caramel and oak. A bit thin, but spicy. A little fine leather, tobacco and fruit.
Finish: Short, but spicy with oak.

Susan

Nose: Caramel and marzipan with vanilla and cherries.
Taste: Vanilla and cherries with nuts and nutmeg.
Finish: Short, warm and spicy.

Notes

Priced to be a reliable everyday house bourbon.

My Score

Notes: ______________________________

____

____

Appearance (1-5) ____ Nose (1-25) ____
Taste (1-25) ____ Finish (1-25) ____
Complexity (1-10) ____ Overall Impression (1-10) ____
SCORE ____

Very Old Barton 90

Proof: 90
Age: 6 years
Type: Straight Bourbon
Style: Traditional
Mash Bill: BW 6
Color: Light Amber
Price: $

Mike

Nose: Apples and caramel with fine leather. Honey and sweet spices (nutmeg?). Hint of oak.
Taste: Honey and apples with some cinnamon spice, tobacco and oak.
Finish: Short and spicy. Starts sweet, but gets some peppery spice and oak.

Susan

Nose: Banana nut bread.
Taste: Vanilla, honey and sweet spices including cinnamon and allspice. Fruit on mid-palate.
Finish: Medium and smooth, pepper but no bite.

Notes

Excellent price point on this sipping bourbon that can be enjoyed with a little water. Would make a lovely Manhattan.

My Score

Notes: ________________________________

____

____

Appearance (1-5) ____ Nose (1-25) ____
Taste (1-25) ____ Finish (1-25) ____
Complexity (1-10) ____ Overall Impression (1-10) ____
SCORE ____

Very Old Barton 100

Proof:	100
Age:	6 years
Type:	Straight Bourbon
Style:	Traditional
Mash Bill:	BW 6
Color:	Amber
Price:	$

Mike

Nose: Honey, apples, caramel and fine leather. Sweet spice and vanilla.

Taste: Very smooth with no burn, but a little warmth in the throat. Apples, caramel and tobacco. Sweet spice – cinnamon.

Finish: Long – starts fruity and sweet but gets dry and spicy. The oak really comes through at the end.

Susan

Nose: Black cherry, vanilla bean and cocoa.

Taste: Crème caramel, cherry pie, cinnamon and nuts.

Finish: Smooth, long, warm and sweet with a fine spicy oak ending.

Notes

A lot of complexity for the price. Good neat or with a splash of water. Mike's favorite bourbon to cook with "because its flavors really come through in cooking."

My Score

Notes: ______________________________

____

____

Appearance (1-5) ____ Nose (1-25) ____

Taste (1-25) ____ Finish (1-25) ____

Complexity (1-10) ____ Overall Impression (1-10) ____

SCORE ____

Zackariah Harris

Proof: 80
Age: NAS
Type: Straight Bourbon
Style: Traditional
Mash Bill: BW 4
Color: Light Straw
Price: $

Mike

Nose: Vanilla and cherries with a little leather and oak.
Taste: Light and sweet – Corn, vanilla and a hint of pitted fruit; apricots or cherries
Finish: Surprisingly nice with sweet fruit and peppery on the end.

Susan

Nose: Banana, candy corn, and oak.
Taste: Sweet corn, touch of clove. Vanilla predominates on the mid-palate.
Finish: Unexpectedly long for its proof, with sweet spices at the end.

Notes

A rather more complex sip than might be expected from the proof and the price.

My Score

Notes: ____

____

____

Appearance (1-5) ____ Nose (1-25) ____
Taste (1-25) ____ Finish (1-25) ____
Complexity (1-10) ____ Overall Impression (1-10) ____

SCORE ____

Barton 1792 Additional Releases

Name: ____________________

Distillery: ____________________

Proof: ________ **Mash Bill:** ________
Age: ________ ________
Type: ________ **Color:** ________
Style: ________ **Price:** ________

My Score

Notes: ____________________
____
____

Appearance (1-5) ___ Nose (1-25) ___
Taste (1-25) ___ Finish (1-25) ___
Complexity (1-10) ___ Overall Impression (1-10) ___
SCORE ___

Name: ____________________

Distillery: ____________________

Proof: ________ **Mash Bill:** ________
Age: ________ ________
Type: ________ **Color:** ________
Style: ________ **Price:** ________

My Score

Notes: ____________________
____
____

Appearance (1-5) ___ Nose (1-25) ___
Taste (1-25) ___ Finish (1-25) ___
Complexity (1-10) ___ Overall Impression (1-10) ___
SCORE ___

Buffalo Trace

www.buffalotracedistillery.com

Limited Bottlings:

- E.H. Taylor Special Editions
- Single Oak Project
- Other Experimental Bottlings

Whiskey may very well have been made on the site of the current distillery as early as 1775. Storied owners and operators over the intervening years have included E. H. Taylor, George T. Stagg, and Albert Blanton. It was renamed Buffalo Trace in 199 under Sazerac's ownership. (Historically, the low lying location on the banks of the Kentucky river was a bison crossing.) Two mash bills — #1 and #2, which has higher rye content than #1. Exact recipes not available. According to Modern Thirst, #1 is 75% corn, 10% rye, 15% malted barley. (We have also seen accounts claiming it is 80% corn.) Mash bill #2, again from Modern Thirst, may be 75% corn, 15% rye, and 10% malted barley. The Van Winkle bourbons may be 70% corn, 16% wheat, and 14% malted barley. In any event, Buffalo Trace is not telling.

Ancient Age

Proof: 80
Age: NAS
Type: Straight Bourbon
Style: Traditional
Mash Bill: #2
Color: Straw
Price: $

Mike

Nose: Candy corn and sweet spice. Nutmeg? But, simple overall.
Taste: Very thin on the palate. Corn and vanilla with a hint of oak.
Finish: Starts sweet but becomes dry with oak.

Susan

Nose: Sweet licorice and candy corn.
Taste: Orange note evolves into vanilla.
Finish: Medium. Warm without burn.

Notes

Balanced, inexpensive bourbon you may keep on hand for everyday uses.

My Score

Notes: ____
____
____

Appearance (1-5) ____ Nose (1-25) ____
Taste (1-25) ____ Finish (1-25) ____
Complexity (1-10) ____ Overall Impression (1-10) ____
SCORE ____

Ancient Age 10 Star

Proof: 90
Age: NAS
Type: Straight Bourbon
Style: Traditional
Mash Bill: #2
Color: Dark Straw
Price: $

Mike

Nose: Sweet caramel and cherries with a hint of sweet spices – cinnamon and a hint of oak.
Taste: Apricots and dates, caramel and spice – pepper and nutmeg.
Finish: Long and dry with oak tannins and spice.

Susan

Nose: Vanilla ice cream, ripe stone fruit, sweet spices.
Taste: Vanilla with sweet oak tannin, peach and Pear with sprinkling of spice.
Finish: Medium long. Rather hot and peppery on the tongue, but nutmeg revealed at the very end.

Notes

This is a more complex sip than you would expect from its price. Certainly there are more expensive bourbons that are not as interesting.

My Score

Notes: ____

____

____

Appearance (1-5) ____ Nose (1-25) ____
Taste (1-25) ____ Finish (1-25) ____
Complexity (1-10) ____ Overall Impression (1-10) ____

SCORE ____

Ancient Age 90

Proof: 90
Age: 36 months
Type: Straight Bourbon
Style: Traditional
Mash Bill: #2
Color: Straw
Price: $

Mike

Nose: Sweet corn and vanilla, with a hint of oak wood and young ethanol.
Taste: Ethanol and corn; a little vanilla and pepper spice.
Finish: Short and sweet with nutmeg and a hint of oak.

Susan

Nose: Toffee/caramel with notes of sweet smoke, bright berries.
Taste: Sweet caramel and corn. A little clove heat, but no burn.
Finish: Fairly short. Spice smooths to corn and vanilla sweetness.

Notes

While still fairly uncomplicated, the 90 proof, as would be expected, has more heft than the 80 proof Ancient Age.

My Score

Notes: ______________________________

____

____

Appearance (1-5) ____ Nose (1-25) ____
Taste (1-25) ____ Finish (1-25) ____
Complexity (1-10) ____ Overall Impression (1-10) ____

SCORE ____

Ancient Ancient Age

Proof: 86
Age: 10 years
Type: Straight Bourbon
Style: Extra Aged
Mash Bill: #2
Color: Light Amber
Price: $

Mike

Nose: Caramel and fruit – apples, pears, dates. Hint of oak wood and sweet spices.

Taste: Apples and caramel with hint of pecan. Almost chewy mouthfeel with some sweet spice – nutmeg and cinnamon.

Finish: Long. Starts sweet and spicy but gets drier as it progresses.

Susan

Nose: Loads of fruits and spices – peach, pear cinnamon and a little clove and vanilla.

Taste: Peaches and pears dusted in cinnamon. Rich vanilla on mid-palate with a bit of cocoa.

Finish: Long, sweet, and warm. Spice at the very end.

Notes

Beautifully balanced 10-year old bourbon that tastes far more expensive than it is. Very suitable for sipping neat or with a splash.

My Score

Notes: ____________________________________

____

____

Appearance (1-5) ____ Nose (1-25) ____
Taste (1-25) ____ Finish (1-25) ____
Complexity (1-10) ____ Overall Impression (1-10) ____

SCORE ____

Benchmark Bonded

Proof: 100
Age: 4 years
Type: Straight
Style: Bottled-in-Bond
Mash Bill: BT #1
Color: Light Straw
Price: $

Mike

Nose: Caramel and fruit – apricots and peaches – with a hint of nutmeg and oak.
Taste: Caramel and peaches with cinnamon, nutmeg, and oak.
Finish: Long and spicy with oak and cinnamon leading to white pepper.

Susan

Nose: Light caramel and a little stone fruit.
Taste: Caramel and apricots with a dash of baking spice and a dusting of cocoa.
Finish: Lingering and spicy with a touch of burnt sugar.

Notes

Great value at about $20 a bottle. Tastes more expensive.

My Score

Notes: ____

____

____

Appearance (1-5) ____ Nose (1-25) ____
Taste (1-25) ____ Finish (1-25) ____
Complexity (1-10) ____ Overall Impression (1-10) ____
SCORE ____

Benchmark Full Proof

Proof: 125
Age: NAS
Type: Straight
Style: Traditional
Mash Bill: BT #1
Color: Straw
Price: $

Mike

Nose: Caramel and dates, raisins and a little baking spice and oak.
Taste: Caramel, dates, and plums with allspice and oak.
Finish: Long with oak and spice.

Susan

Nose: Loads of caramel with apples, pears, and plums, plus baking spices.
Taste: Cinnamon and caramel. Very warming in the mouth. Fruit plays around the edges.
Finish: Long with peppery warmth that lingers in the throat.

Notes

Solid sip. Water amplifies the fruit.

My Score

Notes: ____
____
____

Appearance (1-5) ____ Nose (1-25) ____
Taste (1-25) ____ Finish (1-25) ____
Complexity (1-10) ____ Overall Impression (1-10) ____
SCORE ____

Benchmark Old No. 8

Proof: 80
Age: NAS
Type: Straight Bourbon
Style: Traditional
Mash Bill: BT #1
Color: Straw
Price: $

Mike

Nose: Vanilla and corn, very light nose. Not complex.
Taste: Very light. Corn and vanilla with some fruit – apples and peaches and sweet spices.
Finish: Very short and sweet with a bit of honey and very little oak.

Susan

Nose: Vanilla, candy corn, whiff of nutmeg and a faint citrus note. (Orange?)
Taste: Vanilla and a little cocoa with new leather and nutty caramel. Surprisingly creamy mouthfeel.
Finish: Medium and sweet. Very smooth.

Notes

A bargain-priced bourbon that would work very well in cocktails. You may even enjoy it neat.

My Score

Notes: ______________________________
____
____

Appearance (1-5) ____ Nose (1-25) ____
Taste (1-25) ____ Finish (1-25) ____
Complexity (1-10) ____ Overall Impression (1-10) ____
SCORE ____

Benchmark Single Barrel

Proof:	95
Age:	NAS
Type:	Straight
Style:	Traditional
Mash Bill:	BT #1
Color:	Pale Straw
Price:	$

Mike

Nose: Caramel and apple with a little oak.
Taste: Vanilla and apple with some pepper spice and oak.
Finish: Long with oak and pepper.

Susan

Nose: Fruity and sweet with some sweet corn and caramel.
Taste: Vanilla, caramel, spice, and a little fruit.
Finish: Warm and lightly tannic.

Notes

Not as fruity as some of the other Benchmark expressions. Try it in a Manhattan where it will pair well with sweet vermouth.

My Score

Notes: ____________________________________

____

____

Appearance (1-5) ____ Nose (1-25) ____
Taste (1-25) ____ Finish (1-25) ____
Complexity (1-10) ____ Overall Impression (1-10) ____

SCORE ____

Benchmark Top Floor

Proof: 86
Age: NAS
Type: Straight
Style: Traditional
Mash Bill: BT #1
Color: Straw
Price: $

Mike

Nose: Corn, vanilla, hint of pears, baking spice, oak.
Taste: Corn, vanilla, pears, cinnamon. and oak.
Finish: Medium long with oak and cinnamon.

Susan

Nose: Caramel, some chocolate, light cinnamon.
Taste: Caramel, cinnamon, some cherries and a little oak.
Finish: Medium long and drying to oak.

Notes

A very nice, inexpensive pour. Good for cocktails, too.

My Score

Notes: ____________________________________

____

____

Appearance (1-5) ____ Nose (1-25) ____
Taste (1-25) ____ Finish (1-25) ____
Complexity (1-10) ____ Overall Impression (1-10) ____

SCORE ____

Blanton's Single Barrel

Proof:	93
Age:	NAS
Type:	Straight Bourbon
Style:	Single Barrel
Mash Bill:	Buffalo Trace #2 which is high in rye
Color:	Amber
Price:	$$$

Mike

Nose: Apples and caramel with a bit of oak and leather. Opens after a minute to reveal cherries and dates.

Taste: Apples, vanilla, oak and leather with some sweet spice – nutmeg.

Finish: Long, dry and spicy, with oak tannins and sweet spices.

Susan

Nose: Honey vanilla, light spiciness with a touch of caramel. Fruit appears late.

Taste: Light, creamy mouthfeel, rich vanilla with prominent notes of apple and sweet spices.

Finish: Lingering and smooth with mellow beeswax at the end.

Notes

First commercially available, post-Prohibition, single barrel bourbon. It was the brainchild of Master Distiller Elmer T. Lee and named in honor of Col. Albert Blanton, distillery manager in the early 20th century.

My Score

Notes: ______________________________

____

____

Appearance (1-5) ____ Nose (1-25) ____

Taste (1-25) ____ Finish (1-25) ____

Complexity (1-10) ____ Overall Impression (1-10) ____

SCORE ____

Blanton's Special Reserve

Proof: 80
Age: NAS
Type: Straight Bourbon, Export Only
Style: Single Barrel, #124
Mash Bill: Buffalo Trace #2 which is high in rye
Color: Amber
Price: $$$

Mike

Nose: Caramel and fruit – apricots? – and a hint of oak. Very simple nose.
Taste: Very light taste. A hint of vanilla and caramel with some oak and leather. A bit thin, but some floral notes late.
Finish: Long – starts sweet and adds spiciness.

Susan

Nose: Very light, simple nose. Whiffs of vanilla and caramel.
Taste: Sweet corn, touch of honey, pleasantly oily mouthfeel. But light. No mistaking the low proof.
Finish: Smooth, long finish that retains sweetness to the end.

Notes

One of several expressions available overseas only. Given the low proof, interesting to have in order to compare to other expression.

My Score

Notes: ______________________________

____

____

Appearance (1-5) ____ Nose (1-25) ____
Taste (1-25) ____ Finish (1-25) ____
Complexity (1-10) ____ Overall Impression (1-10) ____

SCORE ____

Blanton's Export

Proof: 93.9
Age: NAS
Type: Straight Bourbon, Export Only
Style: Single Barrel, #255
Mash Bill: Buffalo Trace #2 which is high in rye
Color: Amber
Price: $$$

Mike

Nose: Fruity. Apples and peaches with vanilla
Taste: Not overly complex. Apples and caramel with some peppery spice and fine leather.
Finish: Long and dry with oak tannins.

Susan

Nose: Sweet caramel apples, but light.
Taste: Creamy mouthfeel, saddle leather and vanilla.
Finish: Long, stays sweet most of the way But has a crisp ending.

Notes

One of several expressions available overseas only. Higher proof than the domestic expression.

My Score

Notes: ______________________________
____
____

Appearance (1-5) ____ Nose (1-25) ____
Taste (1-25) ____ Finish (1-25) ____
Complexity (1-10) ____ Overall Impression (1-10) ____
SCORE ____

Blanton's Gold Edition

Proof: 103
Age: NAS
Type: Straight Bourbon, Export Only
Style: Single Barrel, #157
Mash Bill: Buffalo Trace #2 which is high in rye
Color: Light Amber
Price: $$$

Mike

Nose: Surprisingly simple for its 103 proof. Apples and vanilla with a hint of oak. Drop of water opens it and reveals spice.
Taste: Oak and caramel with a hint of fruit-apple. No alcohol burn, but nice and warm in the throat. Water makes it a bit more fruity.
Finish: Dry with oak tannins and pepper spice, but quickly ends. Water makes it last longer.

Susan

Nose: Light spice, touch of vanilla. Drop of water brings out apple and caramel.
Taste: Vanilla, cherries, barest suggestion of cinnamon. Water reveals notes of chocolate.
Finish: Short and dry with a touch of pepper at the end.

Notes

Available overseas only. Notably higher proof than the domestic expression makes it worth seeking out.

My Score

Notes: ______________________________

____

____

Appearance (1-5) ____ Nose (1-25) ____
Taste (1-25) ____ Finish (1-25) ____
Complexity (1-10) ____ Overall Impression (1-10) ____

SCORE ____

Buffalo Trace

Proof: 90
Age: NAS
Type: Straight Bourbon
Style: Traditional
Mash Bill: BT #1
Color: Amber
Price: $$

Mike

Nose: Apples and pears with sweet spices. Allspice, oak and honey with fine leather.

Taste: Honey and vanilla with some ripe pears. A hint of spice and oak wood. Nice Mouthfeel with no burn and little heat.

Finish: Short and sweet with spice and fruit. Second sip brings out the spice.

Susan

Nose: Strong caramel backed up with new leather, sweet spices and ripe apples.

Taste: Rich vanilla/caramel with dark fruit, honey, leather, tobacco and black pepper.

Finish: Medium. Warm and spicy.

Notes

Very versatile bourbon. Sip neat, on the rocks, or as the basis for a fine cocktail. (Susan loves this in a Manhattan.) Would good well at dinner with steak or even fish.

My Score

Notes: ____________________

____

____

Appearance (1-5) ____ Nose (1-25) ____
Taste (1-25) ____ Finish (1-25) ____
Complexity (1-10) ____ Overall Impression (1-10) ____

SCORE ____

Col. E.H. Taylor, Jr. Barrel Proof

Proof: 127.2
Age: NAS
Type: Straight
Style: Barrel Proof, Small Batch
Mash Bill: High Rye
Color: Light Amber
Price: $$$

Mike

Nose: Corn, ripe pear, and baking spices.
Taste: Caramel and pears with nutmeg.
Finish: Spicy and dry with oak.

Susan

Nose: Crème brulee and sweet spices. Water reveals ripe pears and almonds.
Taste: Spicy fruit cake. Add water for caramel, vanilla, fruit, nuts. Very layered.
Finish: Long and warm. Very smooth for the proof!

Notes

A very big bourbon. Uncut and unfiltered. This was Batch 8, Bottle 33.

My Score

Notes: ______________________________________

____

____

Appearance (1-5) ____ Nose (1-25) ____
Taste (1-25) ____ Finish (1-25) ____
Complexity (1-10) ____ Overall Impression (1-10) ____

SCORE ____

Col. E.H. Taylor, Jr. Bottled-in-Bond

Proof: 100
Age: NAS
Type: Straight Bourbon
Style: Small Batch
Mash Bill: Buffalo Trace #1
Color: Light Amber
Price: $$$

Mike

Nose: Caramel and honey with ripe apple and oak. Water intensifies the nose – more sweet honey.
Taste: Honey and molasses with apple fruit and pepper spice. A bit of tobacco and oak. Water brings out the fruit.
Finish: Very nice, with dry oak at the end, though water shortens it some.

Susan

Nose: Vanilla and roasted corn. Plus nutmeg and a little fruit and honey.
Taste: Rich mouthfeel. Hazelnuts, vanilla, dried apples. Water reveals honey, sweet spices, and some chocolate.
Finish: Long and spicy. Goes on and on. Touch of chocolate at the very end.

Notes

Col. E.H. Taylor, former owner of what is now the Buffalo Trace Distillery, promoted the Bottled-in-Bond Act of 1897 to guarantee bourbon quality. Annual releases of different expressions of Col. E.H. Taylor honor his legacy.

My Score

Notes: ________________________________

____

____

Appearance (1-5) ____ Nose (1-25) ____
Taste (1-25) ____ Finish (1-25) ____
Complexity (1-10) ____ Overall Impression (1-10) ____

SCORE ____

Col. E.H. Taylor, Jr. Single Barrel

Proof:	100
Age:	At least 4 years
Type:	Straight
Style:	Single Barrel, Bottled-in-Bond
Mash Bill:	High Rye
Color:	Light Amber
Price:	$$$

Mike

Nose: Caramel corn, ripe apple, some floral notes.
Taste: Caramel, apple, baking spice, and oak pepper spice.
Finish: Dry oak tannins and pepper spice.

Susan

Nose: Caramel, ripe apples, walnuts, sweet mint.
Taste: Caramel/vanilla/chocolate sundae! Water brings out the fruit.
Finish: Sweet oak drying to spices.

Notes

Another big, layered sip from the Taylor series. Well worth seeking out.

My Score

Notes: ____

____

____

Appearance (1-5)	____	Nose (1-25)	____
Taste (1-25)	____	Finish (1-25)	____
Complexity (1-10)	____	Overall Impression (1-10)	____

SCORE ____

Eagle Rare Single Barrel

Proof: 90
Age: 10 years
Type: Straight Bourbon
Style: Single Barrel, Extra Aged
Mash Bill: Buffalo Trace #1
Color: Amber
Price: $$$

Mike

Nose: Caramel and vanilla with a hint of citrus fruit. Some sweet spices and oak and fine leather.
Taste: Caramel and fruit – citrus and dates, with some honey and oak.
Finish: Long and dry with oak tannins with Some sweet spices and vanilla.

Susan

Nose: Vanilla, mint, new leather, and a trace of cocoa. Layers of sweet spice (cinnamon) and orange peel.
Taste: Caramel, toffee. Fine, chewy mouthfeel. Apple and stone fruit with dash of cinnamon.
Finish: Long and peppery with surprising sweet caramel at the very end.

Notes

A premium single barrel that is well worth its price. Its complexity merits multiple samplings. Sip neat or with a small splash or water.

My Score

Notes: ____

____

____

Appearance (1-5) ____ Nose (1-25) ____
Taste (1-25) ____ Finish (1-25) ____
Complexity (1-10) ____ Overall Impression (1-10) ____

SCORE ____

Eagle Rare 17-Year-Old

Proof: 90
Age: 17 years
Type: Straight Bourbon
Style: Single Barrel, Extra Aged
Mash Bill: Buffalo Trace #1
Color: Amber Red
Price: $$$$

Mike

Nose: Dominated by the barrel – oak, wood and leather. Caramel and dark fruit – berries, apricots, dates.

Taste: Cognac-like. Fruity with caramel. Apricots and apples with some fine fine tobacco and a hint of nutmeg.

Finish: Long and fruity sweet, but a bit of dry oak and tobacco.

Susan

Nose: Big vanilla and caramel with nuts and dark fruit. Lots of sweet oak.

Taste: Nutty caramel. A Snickers bar in a glass but with layers of dark fruit and dry oak.

Finish: Very long, lingering and peppery. Very dry at the end.

Notes

Deserving of utmost respect. A great bourbon to sip with a bite of cheese or fruit. Annual releases are produced in very limited quantities.

My Score

Notes: ______________________________

____

____

Appearance (1-5) ____ Nose (1-25) ____
Taste (1-25) ____ Finish (1-25) ____
Complexity (1-10) ____ Overall Impression (1-10) ____

SCORE ____

Elmer T. Lee

Proof: 90
Age: NAS
Type: Straight Bourbon
Style: Single Barrel
Mash Bill: Buffalo Trace #2, which is high in rye
Color: Light Amber
Price: $$

Mike

Nose: Apples and caramel with a bit of oak and leather. Opens after a minute to reveal cherries and dates.
Taste: Apples, vanilla, oak and leather with some sweet spice – nutmeg.
Finish: Long, dry and spicy, with oak tannins and sweet spices.

Susan

Nose: Honey vanilla, light spiciness with a touch of caramel. Fruit appears late.
Taste: Light, creamy mouthfeel, rich vanilla with prominent notes of apple and sweet spices.
Finish: Lingering and smooth with mellow beeswax at the end.

Notes

Elegant, medium-bodied bourbon that would please both new and experienced bourbon sippers. Each barrel for bottling was hand selected by the late Master Distiller Emeritus Elmer T. Lee.

My Score

Notes: ______________________________

____

____

Appearance (1-5) ____ Nose (1-25) ____
Taste (1-25) ____ Finish (1-25) ____
Complexity (1-10) ____ Overall Impression (1-10) ____

SCORE ____

George T. Stagg

Proof:	142.6
Age:	NAS, but more than 10 years
Type:	Straight
Style:	Small Batch, Extra Aged, Unchilled, Unfiltered
Mash Bill:	Buffalo Trace #1
Color:	Bronze
Price:	$$$$

Mike

Nose: Alcohol. Caramel and oak, fine leather and tobacco with pitted fruit – apricots and dates. A little spice – sweet cinnamon, nutmeg and allspice. Splash of water brings out oak and pecans.

Taste: Sweet fruit, almost brandy-like, with caramel and nutmeg. A splash of water brings out complexity with fruit, spice and oak.

Finish: Short and sweet. A splash of water makes the finish last longer and a bit drier.

Susan

Nose: Nutty, rich, and sweet. Lots of oak and spice and fruit layers. Strong note of beeswax.

Taste: Concentrated vanilla and oak work along the tongue revealing stone fruit, cinnamon and honey.

Finish: Dry, short and surprisingly light. Some pepper and spice.

Notes

A barrel-proof bourbon of great complexity when a few drops of water are added. Very limited annual release.

My Score

Notes: ______________________________

____

____

Appearance (1-5) ____ Nose (1-25) ____

Taste (1-25) ____ Finish (1-25) ____

Complexity (1-10) ____ Overall Impression (1-10) ____

SCORE ____

Hancock's President's Reserve

Proof: 88.9
Age: NAS
Type: Straight Bourbon
Style: Single Barrel
Mash Bill: BT #2
Color: Dark Straw
Price: $$$

Mike

Nose: Vanilla and spice – nutmeg and allspice. Fine leather and oak. Light fruit could be pears or very ripe apples.

Taste: Nice mouthfeel. Apples and vanilla with sweet spices – cinnamon and nutmeg.

Finish: Long and fruity sweet. Almost brandy-like with some spicy oak tannins.

Susan

Nose: Honeyed caramel, brown sugar, light oak and sweet spices.

Taste: Vanilla, almonds and baked apples with a sprinkling of nutmeg and cinnamon.

Finish: Long, warm and sweet with honey returning at the end.

Notes

For drinkers who favor the fruity, spicy style of bourbon, this will be one to have for special occasions. Sip it neat or with a single ice cube.

My Score

Notes: ____

____

____

Appearance (1-5) ____ Nose (1-25) ____
Taste (1-25) ____ Finish (1-25) ____
Complexity (1-10) ____ Overall Impression (1-10) ____

SCORE ____

Old Charter 8-Year-Old

Proof: 80
Age: 8 years
Type: Straight Bourbon
Style: Traditional
Mash Bill: Buffalo Trace #1
Color: Dark Straw
Price: $

Mike

Nose: Caramel corn and vanilla with a bit of oak. Very simple nose.

Taste: Sweet corn and vanilla with just a hint of Corn sweetness. Simple and straightforward. spice. A bit thin in the mouth.

Finish: Short and dry with oak tannins.

Susan

Nose: Light oak, vanilla, faint whiff of mint.

Taste: Oaky. Vanilla predominant with a bit of

Finish: Short and warm. Quite dry. No bite.

Notes

A good simple bourbon that goes well with food. You certainly can't beat the price.

My Score

Notes: ______________________________

____

____

Appearance (1-5) ____ Nose (1-25) ____

Taste (1-25) ____ Finish (1-25) ____

Complexity (1-10) ____ Overall Impression (1-10) ____

SCORE ____

Old Charter 10-Year-Old

Proof: 86
Age: 10 years
Type: Straight Bourbon
Style: Extra Aged
Mash Bill: Buffalo Trace #1
Color: Dark Straw
Price: $

Mike

Nose: Caramel corn and sweet spice with apple fruit and oak.
Taste: Corn and vanilla with sweet spices – cinnamon and nutmeg. Apples and pears with oak tannins. Great moth feel.
Finish: Long and spicy dry.

Susan

Nose: Sweet vanilla, with subtle touches of honey and caramel.
Taste: Vanilla and spiced pear. Not too sweet. Not very complex, but very enjoyable.
Finish: Medium long. Starts sweet, turns dry, but hen ends with a touch of corn sweetness.

Notes

Enjoyable on its own, but would also has the character to make an excellent bourbon cocktail ingredient.

My Score

Notes: ____

____

____

Appearance (1-5) ____ Nose (1-25) ____
Taste (1-25) ____ Finish (1-25) ____
Complexity (1-10) ____ Overall Impression (1-10) ____
SCORE ____

Old Charter 101

Proof: 101
Age: NAS
Type: Straight Bourbon
Style: Traditional
Mash Bill: Buffalo Trace #1
Color: Light Straw
Price: $

Mike

Nose: Sweet caramel and vanilla. Not overly complex. Just a hint of oak and leather
Taste: Sweet corn and vanilla with some sweet spices and oak. A whisper of fruit – apples and cherries.
Finish: Long and sweet with caramel and spice.

Susan

Nose: Caramel, oranges and a note of honey. Pretty simple.
Taste: Oak tannins, vanilla, spices. There is some orchard fruit lurking in the background.
Finish: Medium long. Flash of sweetness before a smooth, dry ending.

Notes

A great bourbon to sip with chocolate. Relatively high proof means it will hold up well in a bourbon and soda.

My Score

Notes: ______________________________

____

____

Appearance (1-5) ____ Nose (1-25) ____
Taste (1-25) ____ Finish (1-25) ____
Complexity (1-10) ____ Overall Impression (1-10) ____

SCORE ____

Old Rip Van Winkle 90°

Proof: 90
Age: 10 years
Type: Straight
Style: Wheated, Extra Aged
Mash Bill: Proprietary, but wheat in place of rye
Color: Pale Amber
Price: $$$

Mike

Nose: Caramel and vanilla with some fine leather, apple/pear fruit and pecans. Not overly complex.
Taste: Rich caramel with apple fruit and pecans. Again, not overly complex.
Finish: Nice and long with dry oak and a hint of spice and nutmeg.

Susan

Nose: Light notes of vanilla, spice, and apple. Rather simple.
Taste: Sweet caramel with underlying soft nuttiness. Surprisingly simple.
Finish: Smooth with a spicy little kick at the end.

Notes

No longer part of the Van Winkle portfolio, but the big whiskey bars may have some. For a price.

My Score

Notes: ____________________________________

____

____

Appearance (1-5) ____ Nose (1-25) ____
Taste (1-25) ____ Finish (1-25) ____
Complexity (1-10) ____ Overall Impression (1-10) ____
SCORE ____

Old Rip Van Winkle 107°

Proof: 107
Age: 10 years
Type: Straight
Style: Wheated, Extra Aged
Mash Bill: Proprietary, but wheat in place of rye
Color: Dark Amber
Price: $$$

Mike

Nose: Caramel and vanilla with some fine leather and tobacco. Ripe apple/pear with hazelnuts and oak.
Taste: Oak with caramel and apple. Also nutmeg and allspice.
Finish: Sweet with caramel apple and spice. The oak is hiding in the background, But not dry oakiness.

Susan

Nose: Vanilla, caramel apple, and light licorice.
Taste: Tobacco, dark cherry and light maple syrup. with faint spicy notes.
Finish: Vanilla and caramel corn, with touch of oak at the end, but the finish fades rather more quickly than one might want.

Notes

A very satisfying sip. The higher proof intensifies the flavors also present in the 10 year, 90 proof. Fascinating to compare the two side by side since aging and mash bill are the same, only the proof is different.

My Score

Notes: ____

____

____

Appendix (1-5) ____ Nose (1-25) ____
Taste (1-25) ____ Finish (1-25) ____
Complexity (1-10) ____ Overall Impression (1-10) ____

SCORE ____

Old Taylor

Proof:	80
Age:	NAS
Type:	Straight Bourbon
Style:	Traditional
Mash Bill:	Buffalo Trace #1
Color:	Light Straw
Price:	$

Mike

Nose: Very simple nose with corn and vanilla.
Taste: Very light taste. Corn and vanilla with a little pepper spice.
Finish: Longer than expected with wood tannins giving a dry finish.

Susan

Nose: Spice leaps out followed by vanilla and a drop of honey.
Taste: Warm vanilla with caramel and pleasant sweet spices. Light mouthfeel.
Finish: Medium finish ends with some heat.

Notes

Uncomplicated bourbon bottled a low proof. The price is such that a bourbon lover will want to have it on hand if, for no other reason, to honor the memory of its namesake, a giant of the bourbon industry who was instrumental in the Bottled in Bond Act.

My Score

Notes: ______________________________

____

____

Appearance (1-5) ____ Nose (1-25) ____
Taste (1-25) ____ Finish (1-25) ____
Complexity (1-10) ____ Overall Impression (1-10) ____

SCORE ____

Pappy Van Winkle's Family Reserve 15-Year-Old

Proof: 107
Age: 15 years
Type: Straight
Style: Wheated, Extra Aged
Mash Bill: Proprietary, but wheat in place of rye
Color: Deep Amber Red
Price: $$$$

Mike

Nose: Green apple and vanilla with fine tobacco and oak.
Taste: Caramel and fruit – pears! With oak tannins and pecans.
Finish: Long and dry with oak tannins and anise.

Susan

Nose: Big caramel right at the front bolstered by apple, pear and rich vanilla and underlying oak and dark honey.
Taste: Very different from the nose showing more vanilla, hint of cherry and black pepper.
Finish: Long smooth finish of polished oak.

Notes

Satisfyingly complex sip. A splash of water could bring out more complexity and tame the high proof a bit.

My Score

Notes: ____________________________________

____

____

Appearance (1-5) ____ Nose (1-25) ____
Taste (1-25) ____ Finish (1-25) ____
Complexity (1-10) ____ Overall Impression (1-10) ____

SCORE ____

Pappy Van Winkle's Family Reserve 20-Year-Old

Proof: 90.4
Age: 20 years
Type: Straight
Style: Wheated, Extra Aged
Mash Bill: Proprietary, but wheat in place of rye
Color: Red
Price: $$$$

Mike

Nose: Sweet crème brûlée, with fine leather and oak wood. Background dates and cherries.
Taste: Sweet dark fruit and caramel with Tobacco and leathery oak tannins.
Finish: Long and dry with oak tannins and sweet vanilla.

Susan

Nose: Full of burnt sugar, dark fruit and oak.
Taste: Sweet saddle leather, light vanilla and underlying caramel.
Finish: Sweet, smooth vanilla finish. Wood fades to background.

Notes

A bourbon to savor neat. Cradle the glass to warm the whiskey. Pairs well with a cigar.

My Score

Notes: ____

____

____

Appearance (1-5) ____ Nose (1-25) ____
Taste (1-25) ____ Finish (1-25) ____
Complexity (1-10) ____ Overall Impression (1-10) ____

SCORE ____

Pappy Van Winkle's Family Reserve 23-Year-Old

Proof:	95.6
Age:	23 years
Type:	Straight
Style:	Wheated, Extra Aged
Mash Bill:	Proprietary, but wheat in place of rye
Color:	Deep Amber Red
Price:	$$$$

Mike

Nose: Dark fruit, but dominated by oak making it very simple on the nose.
Taste: Dark fruit – plums – with lots of oak.
Finish: Dry oak tannins with some sweet caramel at the end.

Susan

Nose: Masses of oak and dry, dark fruit. (Prunes?)
Taste: Old leather and oak and dark fruit.
Finish: Long, dry finish with hint of sweetness at the end.

Notes

Water will mellow the tannins. Enjoy an expensive serving at a high end whiskey bar, but don't buy into the secondary market madness.

My Score

Notes: ______________________________

____

____

Appearance (1-5) ____ Nose (1-25) ____
Taste (1-25) ____ Finish (1-25) ____
Complexity (1-10) ____ Overall Impression (1-10) ____
SCORE ____

Rock Hill Farms

Proof: 100
Age: NAS
Type: Straight Bourbon
Style: Single Barrel
Mash Bill: BT #2
Color: Amber
Price: $$$

Mike

Nose: Caramel and oak, a hint of milk chocolate. Apples, cherries and some sweet spices.
Taste: Spicy. Pepper and cinnamon with apple fruit and vanilla. Warmth, but no burn. Almost chewy. Water brings out strong vanilla flavor.
Finish: Long and spicy. Slightly dry with oak wood and pepper.

Susan

Nose: Chocolate (!) followed by vanilla and pear. Whiff of candied almonds.
Taste: Chocolate and vanilla with lots of cinnamon, nutmeg and vanilla bean. Notes of saddle leather and tobacco, especially with a little water.
Finish: Long and spicy, with a surprising note of candied orange peel at the very end.

Notes

Very elegant sip. Excellent with chocolate. Named in honor of Col. Albert Blanton's Farm that was next to the current Buffalo Trace distillery site.

My Score

Notes: ______________________________
____
____

Appearance (1-5) ____ Nose (1-25) ____
Taste (1-25) ____ Finish (1-25) ____
Complexity (1-10) ____ Overall Impression (1-10) ____

SCORE ____

Stagg, Jr.

Proof:	134.4
Age:	NAS, but 8-9 years according to distiller
Type:	Straight Bourbon
Style:	Uncut, Unfiltered, Barrel Proof, Small Batch
Mash Bill:	BT #1
Color:	Dark Amber
Price:	$$$

Mike

Nose: Surprisingly light nose. Corn, vanilla, caramel and sweet spices. After 5 minutes, roasted marshmallow, caramel, oak.

Taste: Hot without water, but added water enhances caramel, apricot, nutmeg and cinnamon.

Finish: Pepper spice and oak tannins. Dilute to 50/50 water/bourbon and lingering finish is sweeter with spiced apricots.

Susan

Nose: Crème brûlée, dash of cocoa, and cinnamon. Opens up into complex spices with cherries.

Taste: Chocolate, dark cherry, honeycomb, cloves, cinnamon with vanilla oak.

Finish: Long, lingering. Dark fruit fades to black pepper and oak which lasts and lasts.

Notes

Let this barrel-proof, complex bourbon sit in the glass for a bit and add a little water to release all the layers of flavor. Very, very rewarding with that splash.

My Score

Notes: ______________________________

____

____

Appearance (1-5)	___	Nose (1-25)	___
Taste (1-25)	___	Finish (1-25)	___
Complexity (1-10)	___	Overall Impression (1-10)	___

SCORE ___

Van Winkle Special Reserve

Proof: 90.4
Age: 12 years
Type: Straight
Style: Wheated, Extra Aged
Mash Bill: Proprietary, but wheat in place of rye
Color: Amber
Price: $$$

Mike

Nose: Caramel and milk chocolate with hint of nuts and dark Cavendish pipe tobacco.
Taste: Caramel and apples with a hint of oak. Not overly complex. No burn at all.
Finish: Short finish. Sweet caramel, but then gone.

Susan

Nose: Baked apples, spicy vanilla, light caramel.
Taste: Honeyed flavor in the front moving to apple in the middle.
Finish: Very smooth finish but fades quickly.

Notes

Exhibits classic wheated bourbon attributes, especially mellowness and smoothness. Not a complex bourbon, but very enjoyable sipped neat.

My Score

Notes: ______________________________
____
____

Appearance (1-5) ____ Nose (1-25) ____
Taste (1-25) ____ Finish (1-25) ____
Complexity (1-10) ____ Overall Impression (1-10) ____

SCORE ____

Weller 12-Year-Old

Proof: 90
Age: 12 years
Type: Straight Bourbon
Style: Wheated, Extra Aged
Mash Bill: Wheat in place of rye
Color: Straw
Price: $$$

Mike

Nose: Oak and caramel with apple fruit. Very simple nose.

Taste: Sweet caramel and apples, pecans and oak. A nice warmth in the mouth

Finish: Nice and dry with oak, but a little sweet with ripe apples.

Susan

Nose: Vanilla, faint sweet nuttiness with underlying caramel.

Taste: Oak tannins predominate. Some apple but not a lot of depth.

Finish: Medium long with a sprinkle of pepper. Very smooth with no burn at all.

Notes

Extra aging has, not surprisingly, introduced more oak character to this Weller than to some of the other Weller expressions. Also hard to find, since the word is out that this is the bourbon from which Julian Van Winkle selects barrels to become Pappy Van Winkle.

My Score

Notes: ______________________________

____

____

Appearance (1-5) ____ Nose (1-25) ____
Taste (1-25) ____ Finish (1-25) ____
Complexity (1-10) ____ Overall Impression (1-10) ____

SCORE ____

Weller Antique

Proof: 107
Age: NAS
Type: Straight Bourbon
Style: Wheated
Mash Bill: Wheat in place of rye
Color: Light Amber
Price: $$$

Mike

Nose: Apples, pears, vanilla and corn. A hint of oak and fine leather and pecans.
Taste: Nice mouthfeel. Almost chewy. Apples and caramel with hazelnuts and spice.
Finish: Long and spicy with cinnamon and pepper with oak. very good finish.

Susan

Nose: Caramel jumps right out, plus sweet corn and a suggestion of citrus. Quite oaky.
Taste: Begins with caramel sweetness and reveals vanilla and oak. More nutty than fruity with a touch of cinnamon.
Finish: Long, spicy, but not hot.

Notes

Fine example of the wheated bourbon style at a very reasonable (dare we say bargain?) price.

My Score

Notes: ______________________________

____

____

Appearance (1-5) ____ Nose (1-25) ____
Taste (1-25) ____ Finish (1-25) ____
Complexity (1-10) ____ Overall Impression (1-10) ____

SCORE ____

Weller Special Reserve

Proof: 90
Age: NAS
Type: Straight Bourbon
Style: Wheated
Mash Bill: Contains wheat, no rye
Color: Dark Straw
Price: $$

Mike

Nose: Candy – caramel corn and vanilla with some honey and nuts – hazelnuts.
Taste: Sweet corn, apples and pears with some caramel and nut flavor – pecans and hazelnuts. Very nice warmth and mouthfeel.
Finish: Long and dry with oak and spice.

Susan

Nose: Nutty caramel, light oak. a little apple and a whiff of honey.
Taste: More apple on the palate than in the nose. caramel and oak with some dried fruit and a flash of black pepper.
Finish: Short with a dry peppery finish.

Notes

A good bourbon with food or a cigar. Very nice sipped neat.

My Score

Notes: ____________________________________

____

____

Appearance (1-5) ____ Nose (1-25) ____
Taste (1-25) ____ Finish (1-25) ____
Complexity (1-10) ____ Overall Impression (1-10) ____
SCORE ____

William Larue Weller

Proof:	123.4
Age:	Minimum 10 years
Type:	Straight
Style:	Small Batch, Extra Aged, Unchilled, Unfiltered
Mash Bill:	Wheated
Color:	Amber Orange
Price:	$$$$

Mike

Nose: Caramel and vanilla with apples and and pears. A little sweet spice with cinnamon and nutmeg. Oak wood. A little water opens the sweetness.

Taste: For such a high proof, not as much heat as expected. Rich caramel/maple syrup with apples and hazelnuts. Water takes the heat down a notch and sweetens the taste.

Finish: Not as long as expected, but of decent length with oak tannins and sweet fruit. Water makes the finish last longer.

Susan

Nose: Vanilla, candy corn, marzipan and pears. Faintly spicy.

Taste: Oak jumps out in front and settles into dry vanilla, apples, and almonds.

Finish: Medium long. Oak predominates. Dash of black pepper at the very end.

Notes

Part of the Antique Collection along with George T. Stagg, Eagle Rare 17-year-old and two ryes.

My Score

Notes: ______________________________

____

____

Appearance (1-5)	____	Nose (1-25)	____
Taste (1-25)	____	Finish (1-25)	____
Complexity (1-10)	____	Overall Impression (1-10)	____

SCORE ____

Buffalo Trace Additional Releases

Name: ________________________________

Distillery: ________________________________

Proof: ____________ **Mash Bill:** ____________
Age: ____________ ____________
Type: ____________ **Color:** ____________
Style: ____________ **Price:** ____________

My Score

Notes: ________________________________

____

____

Appearance (1-5) ____ Nose (1-25) ____
Taste (1-25) ____ Finish (1-25) ____
Complexity (1-10) ____ Overall Impression (1-10) ____
SCORE ____

Name: ________________________________

Distillery: ________________________________

Proof: ____________ **Mash Bill:** ____________
Age: ____________ ____________
Type: ____________ **Color:** ____________
Style: ____________ **Price:** ____________

My Score

Notes: ________________________________

____

____

Appearance (1-5) ____ Nose (1-25) ____
Taste (1-25) ____ Finish (1-25) ____
Complexity (1-10) ____ Overall Impression (1-10) ____
SCORE ____

Wild Turkey

www.wildturkeybourbon.com

Bourbon Distilleries:

- Wild Turkey Distillery, Lawrenceburg, Ky.

Austin, Nichols and Company was founded in 1855 as a grocery in New York City where it became involved in the food importing and distributing business. After the Repeal of Prohibition the company entered the wine and liquor business. By 1938 it has sold the food side of the business to concentrate on spirits. In 1942 they introduce Wild Turkey Bourbon at 101 proof and 8 years old. In 1969 the tobacco firm of Liggett and Meyers purchases the company. In 1971 they purchase the Ripy distillery in Lawrenceburg, Kentucky. In 1980 Pernod Richard purchases the brand and distillery. The Campari Group purchased the brand and distillery in 2009.

A note on mash bill — Wild Turkey makes no secret about the fact that every one of its bourbons is made with the same mash bill. But they do not say what it is. Modern Thirst claims it is 75% corn, 13% rye, 12% malted barley.

Special Release: Annual release of a single barrel, extra-aged bourbon called Tradition. Age differs each year.

Bond & Lillard

Proof: 100
Age: NAS
Type: Straight Bourbon
Style: Small Batch
Mash Bill: Not available
Color: Light Straw
Price: $$$$

Mike

Nose: A little floral with light fruit – berries and vanilla.
Taste: Dries apricots and dates with a hint of baking spice and tobacco. Brandy-like.
Finish: Long and dry with oak and spice.

Susan

Nose: Sweet mint, ripe apples, vanilla custard.
Taste: Fruity with vanilla and sweet nuts. An elegant sip.
Finish: Stays sweet even through a peppery ending.

Notes

Price is about $50 for a 375ml bottle. Part of the Wild Turkey Whiskey Barons Collection reviving historic brands. Even though there is no age statement on the bottle, the distillery told us the whiskey used is between 6 and 8 years old. This was Batch No. 1.

My Score

Notes: ____

____

____

Appearance (1-5) ____ Nose (1-25) ____
Taste (1-25) ____ Finish (1-25) ____
Complexity (1-10) ____ Overall Impression (1-10) ____
SCORE ____

Old Ripy

Proof: 104
Age: NAS
Type: Straight
Style: Non-Chill Filtered
Mash Bill: Not available
Color: Light Straw
Price: $$$

Mike

Nose: Corn, apples and pears with some caramel.
Taste: Creamy mouthfeel with apples and caramel and baking spices.
Finish: Dry with oak leading into a little pepper spice.

Susan

Nose: Vanilla, sweet corn, almonds.
Taste: Caramel, vanilla. Very smooth at 104 proof. Dash of cinnamon. Splash of water reveals apples and pears.
Finish: Sweet oak. Some spice.

Notes

Price is about $50 for a 375ml bottle. Part of the Wild Turkey Whiskey Barons Collection reviving historic brands. Even though there is no age statement on the bottle, the distillery told us the whiskey used is between 8 and 12 years old.

My Score

Notes: ____________________________________
____
____

Appearance (1-5) ____ Nose (1-25) ____
Taste (1-25) ____ Finish (1-25) ____
Complexity (1-10) ____ Overall Impression (1-10) ____

SCORE ____

Russell's Reserve

Proof: 90
Age: 10 years
Type: Straight
Style: Small Batch
Mash Bill: Not available
Color: Amber
Price: $$

Mike

Nose: Vanilla and honeysuckle with a bit of oak. Very complex – leather and tobacco and sweet spices. Water brings out some very long time.
Taste: Very nice thick and chewy mouthfeel with caramel, oak and sweet spice. A hint of apples and apricots.
Finish: Long. Starts sweet and spicy, but gets a nice oak dryness. Water makes the finish even more intense.

Susan

Nose: Sweet almond (marzipan), vanilla, cocoa, with sweet spices and cherry. Complex. This is a bourbon you can just nose for a nice caramel.
Taste: Sweet caramel with lots of orchard fruits, nuts, and sweet spices. Flash of cocoa, too.
Finish: Long and smooth. The fruit dries to pleasant nuttiness at the end.

Notes

Beautiful, snifter-worthy bourbon to be savored after a fine meal. Just a few drops of water help bring out the many flavor layers.

My Score

Notes: ____

____

____

Appearance (1-5) ____ Nose (1-25) ____
Taste (1-25) ____ Finish (1-25) ____
Complexity (1-10) ____ Overall Impression (1-10) ____

SCORE ____

Russell's Reserve Single Barrel

Proof:	110
Age:	NAS
Type:	Straight
Style:	Single Barrel, Non-Chill Filtered
Mash Bill:	Not available
Color:	Amber
Price:	$$$

Mike

Nose: Very intense. Caramel and sweet spices with hint of chocolate and oak. Water opens up the caramel and fruit.

Taste: Oak tannins and caramel with some fruit – apples and pears.

Finish: Long and spicy with oak tannins.

Susan

Nose: Dark honey, maple syrup, intense caramel, pecans, beeswax, oak and touch of leather.

Taste: Honey, almonds, cocoa, with dried cherries layered throughout the rich caramel.

Finish: Long and smooth starting with fruit and honey and ending with peppery spice.

Notes

Worth every penny of the sticker price. Beautifully complex and balanced.

My Score

Notes: ______________________________

____

____

Appearance (1-5) ____ Nose (1-25) ____

Taste (1-25) ____ Finish (1-25) ____

Complexity (1-10) ____ Overall Impression (1-10) ____

SCORE ____

W.B. Saffell

Proof:	107
Age:	NAS
Type:	Straight
Style:	Traditional
Mash Bill:	Not available
Color:	Pale Straw
Price:	$$$

Mike

Nose: Vanilla and old leather with hints of spice and oak.
Taste: Vanilla, corn, pepper, with a hint of fruit.
Finish: Fairly long with lots of oak and pepper.

Susan

Nose: Honey, vanilla, light fruit.
Taste: Vanilla, sweet spices, faint cherries.
Finish: Warm and peppery.

Notes

Water releases more cinnamon. One of the expressions in Wild Turkey's Whiskey Barons Collection, named for William Butler Saffell who opened a Lawrenceburg distillery in 1889.

My Score

Notes: ______________________________

____

____

Appearance (1-5) ____ Nose (1-25) ____

Taste (1-25) ____ Finish (1-25) ____

Complexity (1-10) ____ Overall Impression (1-10) ____

SCORE ____

Wild Turkey 81

Proof: 81
Age: NAS
Type: Straight
Style: Traditional
Mash Bill: Not available
Color: Pale Straw
Price: $

Mike

Nose: Very light nose. Vanilla and oak with some light fruit and spice. Hard to distinguish the types.

Taste: Nice mouthfeel. Almost chewy. Nice apple fruit and caramel with some tobacco. Better than expected from the nose.

Finish: Short but sweet with vanilla and spices.

Susan

Nose: Light caramel and light citrus with a little vanilla.

Taste: Very smooth. Light caramel corn and a bit of honeyed nuts. Much more interesting than the nose would have indicated.

Finish: Fairly short and sweet with no burn.

Notes

Good introductory bourbon for someone who thinks bourbon is "hot" or "harsh." It would change his/her mind. Refreshing in bourbon and soda and very nice as the base for an Old Fashioned.

My Score

Notes: ______________________________

____

____

Appearance (1-5) ____ Nose (1-25) ____
Taste (1-25) ____ Finish (1-25) ____
Complexity (1-10) ____ Overall Impression (1-10) ____

SCORE ____

Wild Turkey 101

Proof: 101
Age: NAS
Type: Straight
Style: Traditional
Mash Bill: Not available
Color: Light Amber
Price: $

Mike

Nose: Caramel and fine leather. A bit of fruit – cherries and dates – and oak wood.
Taste: Fruit – cherries and berries with caramel and peppery spices. Very full-bodied and flavorful.
Finish: Long and spicy turning to nice, dry oak.

Susan

Nose: Vanilla and pitted fruit with a hint of honey and a whiff of tobacco leaf.
Taste: Lovely creamy mouthfeel. Remarkably smooth for the high proof. Vanilla, honey, caramel corn, citrusy fruit and spices.
Finish: Long smooth, spicy finish ends with flavor of honeycomb.

Notes

An absolute classic. Great sipping bourbon which also goes well with steak or a cigar. Splendid for a hot toddy on a cold winter evening.

My Score

Notes: ______________________________

____

____

Appearance (1-5)	____	Nose (1-25)	____
Taste (1-25)	____	Finish (1-25)	____
Complexity (1-10)	____	Overall Impression (1-10)	____

SCORE ____

Wild Turkey Kentucky Spirit

Proof: 101
Age: NAS
Type: Straight
Style: Single Barrel
Mash Bill: Not available
Color: Amber
Price: $$$

Mike

Nose: Rich caramel and milk chocolate. Fine tobacco and oak with apple/pear fruit. A little water brings out the caramel.

Taste: Apples and sweet spice – nutmeg and cinnamon. With oak tannins and caramel.

Finish: Long and dry with oak tannins and spice. Water tames the dryness, but makes it less interesting.

Susan

Nose: Caramel, green apple, pecans, beeswax, oak and a touch of leather.

Taste: Nuts, apples and honey supplement strong vanilla/caramel flavor. Dashes of cinnamon and sweet chocolate.

Finish: Long and smooth. Absolutely no burn. Dries to a spicy oak ending.

Notes

Enjoy neat. Our particular sample was bottled in May, 2012 from Barrel #41 which had been stored in Warehouse M, Rick 3.

My Score

Notes: ____

____

____

Appearance (1-5) ____ Nose (1-25) ____
Taste (1-25) ____ Finish (1-25) ____
Complexity (1-10) ____ Overall Impression (1-10) ____

SCORE ____

Wild Turkey Longbranch

Proof: 86
Age: NAS
Type: Straight
Style: Traditional
Mash Bill: Not available
Color: Straw
Price: $$

Mike

Nose: Vanilla and oak with a bit of ripe pears and sweet baking spices.
Taste: Mouthfeel a bit thin. Vanilla and caramel with pears and pepper.
Finish: Dry with oak and pepper.

Susan

Nose: Vanilla, apples, new leather, and baking spice.
Taste: Apples with custard, some cinnamon, honey, and a touch of some.
Finish: Long and warm.

Notes

Created in collaboration with actor Matthew McConaughey who signed on as "creative director" from 2018 to 2022. Filtered through mesquite charcoal.

My Score

Notes: ____________________________________
____
____

Appearance (1-5) ____ Nose (1-25) ____
Taste (1-25) ____ Finish (1-25) ____
Complexity (1-10) ____ Overall Impression (1-10) ____
SCORE ____

Wild Turkey Master's Keep

Proof:	86.8
Age:	17 years
Type:	Straight
Style:	Small Batch, Extra Aged
Mash Bill:	Not available
Color:	Amber
Price:	$$$$

Mike

Nose: Berries, caramel, crème brulee, and tobacco.

Taste: Berries and pears with caramel and oak. Nice mouthfeel with chewy texture and no burn.

Finish: Surprisingly sweet for the age. Just a hint of tannins, a little chocolate and oak.

Susan

Nose: Vanilla, honey ripe apples, oak.

Taste: Vanilla and apples with pepper. Substantial mouthfeel.

Finish: Long and warm with a little cocoa and a peppery ending.

Notes

This was small batch number 1. Was moved during aging from a wood warehouse to a stone warehouse, and back to wood. Creation of Eddie Russell.

My Score

Notes: ______________________________

____

____

Appearance (1-5) ____ Nose (1-25) ____

Taste (1-25) ____ Finish (1-25) ____

Complexity (1-10) ____ Overall Impression (1-10) ____

SCORE ____

Wild Turkey Rare Breed

Proof:	108.2
Age:	NAS
Type:	Straight
Style:	Barrel Proof, Small Batch
Mash Bill:	Not available
Color:	Amber
Price:	$$

Mike

Nose: Dark fruit (cherries/dates) with vanilla and caramel, oak and fine leather. A few drops of water brings out the caramel and fruit.

Taste: Fruit and spice. Apricots and nutmeg/allspice with oak tannins and vanilla.

Finish: Long and dry oak. Just a hint of spice. Water makes the finish longer and spicier.

Susan

Nose: Strong honey on the nose, then vanilla, caramel, ripe peach, almonds and chocolate.

Taste: Rich, chewy, creamy mouthfeel. Vanilla, stone fruits, honey, brown sugar, caramel.

Finish: Long and lingering. Fruit gives way to tannins with a touch of sweet tobacco at the very end.

Notes

Sip neat or with a few drops of water. Perfect after dinner treat with a rich chocolate dessert. Would pair well with a mild cigar.

My Score

Notes: ____________________________________

____

____

Appearance (1-5)	___	Nose (1-25)	___
Taste (1-25)	___	Finish (1-25)	___
Complexity (1-10)	___	Overall Impression (1-10)	___

SCORE ___

Wild Turkey Tradition

Proof: 101
Age: 14 years
Type: Straight
Style: Extra Aged, Limited Release
Mash Bill: Not available
Color: Amber
Price: $$$$

Mike

Nose: Apples, pears, caramel, tobacco and a bit of oak and spice. Opens nicely over time with more fruit and spice.
Taste: Berries and pears, maybe some dates, with vanilla and caramel, tobacco and nutmeg.
Finish: Long. Starts with very sweet vanilla and spice and gets oak as it lasts. Brandy-like.

Susan

Nose: Oak, vanilla, apple with cinnamon and spiced, roasted nuts, old book leather.
Taste: Caramel, honey, vanilla, and lots of figs. Seems a bit closed at first.
Finish: Long. Sweet at beginning and dries slowly. Very, very smooth.

Notes

Sweet fruit and spices emerge that are not apparent at first. There will be nutmeg, cinnamon, and chocolate. Drop of water turns this to milk chocolate. Mike calls this a "book bourbon." Pour a glass, settle in with a good book, and take a sip every 15 minutes or so for emerging flavors.

My Score

Notes: ____

____

____

Appearance (1-5)	____	Nose (1-25)	____
Taste (1-25)	____	Finish (1-25)	____
Complexity (1-10)	____	Overall Impression (1-10)	____

SCORE ____

Wild Turkey Additional Releases

Name: ______________________________

Distillery: ______________________________

Proof: ____________ **Mash Bill:** ____________
Age: ____________ ____________
Type: ____________ **Color:** ____________
Style: ____________ **Price:** ____________

My Score

Notes: ______________________________

____

____

Appearance (1-5) ____ Nose (1-25) ____
Taste (1-25) ____ Finish (1-25) ____
Complexity (1-10) ____ Overall Impression (1-10) ____
SCORE ____

Name: ______________________________

Distillery: ______________________________

Proof: ____________ **Mash Bill:** ____________
Age: ____________ ____________
Type: ____________ **Color:** ____________
Style: ____________ **Price:** ____________

My Score

Notes: ______________________________

____

____

Appearance (1-5) ____ Nose (1-25) ____
Taste (1-25) ____ Finish (1-25) ____
Complexity (1-10) ____ Overall Impression (1-10) ____
SCORE ____

Wilderness Trail

www.wildernesstraildistillery.com

The founders of the distillery, which opened in 2013, are Shane Baker (a mechanical engineer) and Pat Heist (a microbiologist) which is a fortuitous combination for people running a distillery. Their first distillery was in downtown Danville with small pot still. It was an outgrowth of the owners' other company, Ferm Solutions, which cultures and sells customized strains of yeast to distilleries and breweries. The distillery is now locate on a 168-acre site on the outskirts of town which utilizes column, as well as the pot still for the sweet mash whiskeys. In 2022, Campari Group acquired 70% control of the distillery.

Wilderness Trail Small Batch

Proof: 100
Age: 5-6 years
Type: Straight
Style: Bonded, Sweet Mash
Mash Bill: 64% corn, 24% rye, 12% malted barley
Color: Light Amber
Price: $$

Mike

Nose: Vanilla, baking spices, apricots, oak wood.
Taste: Vanilla, apricots, nutmeg, white pepper, and oak.
Finish: Medium long with oak, pepper, and lingering vanilla.

Susan

Nose: Ripe apples, vanilla, and a light dusting of baking spices.
Taste: Light brown sugar, red hots, and fruit cocktail.
Finish: Long and spice without being hot.

Notes

A very balanced, versatile sip. Excellent neat, with a small ice cube, or in a cocktail.

My Score

Notes: ____________________________________

____

____

Appearance (1-5) ____ Nose (1-25) ____
Taste (1-25) ____ Finish (1-25) ____
Complexity (1-10) ____ Overall Impression (1-10) ____
SCORE ____

Wilderness Trail Small Batch Wheated

Proof:	100
Age:	5-6 years
Type:	Straight
Style:	Bonded, Wheated, Sweet Mash
Mash Bill:	64% corn, 24% wheat, 12% malted barley
Color:	Straw
Price:	$$$

Mike

Nose: Vanilla, apples, oak wood.
Taste: Vanilla, apples, pepper spice, oak wood.
Finish: Medium long with oak, pepper spice, and oak wood.

Susan

Nose: Light fruit and floral aromas with a little nutmeg and light brown sugar.
Taste: Multigrain cereal with brown sugar and dried apples.
Finish: Long and pleasantly oaky.

Notes

A very good wheat recipe bourbon. Look for 8-year and single barrel expressions, too.

My Score

Notes: ______________________________

____

____

Appearance (1-5) ____ Nose (1-25) ____
Taste (1-25) ____ Finish (1-25) ____
Complexity (1-10) ____ Overall Impression (1-10) ____

SCORE ____

Wilderness Trail Additional Releases

Name: ______________________________

Distillery: ______________________________

Proof: ____________ **Mash Bill:** ____________

Age: ____________ ____________

Type: ____________ **Color:** ____________

Style: ____________ **Price:** ____________

My Score

Notes: ______________________________

____

____

Appearance (1-5) ____ Nose (1-25) ____

Taste (1-25) ____ Finish (1-25) ____

Complexity (1-10) ____ Overall Impression (1-10) ____

SCORE ____

Name: ______________________________

Distillery: ______________________________

Proof: ____________ **Mash Bill:** ____________

Age: ____________ ____________

Type: ____________ **Color:** ____________

Style: ____________ **Price:** ____________

My Score

Notes: ______________________________

____

____

Appearance (1-5) ____ Nose (1-25) ____

Taste (1-25) ____ Finish (1-25) ____

Complexity (1-10) ____ Overall Impression (1-10) ____

SCORE ____

Willett Distillery

www.willettdistillery.com

Bourbon Distilleries:

- Willett Distillery, Bardstown, Ky.

Thompson Willett and his brother John L. Willett founded the Willett Distillery after the repeal of Prohibition. It opened in 1936. By St Patrick's Day 1937, the first barrel of whiskey was rolled into Warehouse A. For a short period of time in the late 1970's the distillery was converted to a fuel alcohol plant.

On December 16, 1972, a Norwegian decanter craftsman, Even G. Kulsveen, married the daughter of Thompson Willett, Martha Harriett Willett. Together, Even & Martha purchased the property in 1984, formed Kentucky Bourbon Distillers, Ltd., DSP KY-78 and began extensive renovations. Even & Martha's children Britt and Drew Kulsveen, joined the family business in the 2000's. After several years of refurbishment, the family began distilling again in 2012. Today Britt is President and Chief Whiskey Officer. Drew serves as Willett Master Distiller.

Johnny Drum Black Label

Proof: 86
Age: 8 years
Type: Straight Bourbon
Style: Traditional
Mash Bill: Proprietary
Color: Dark Amber
Price: $$$

Mike

Nose: Light. Caramel and vanilla with a hint of oak.
Taste: Very lightly flavored. Caramel and vanilla with a hint of honey.
Finish: Very long and nice with a peppery ending and a bit of oak.

Susan

Nose: Vanilla, marzipan, bright herbal notes.
Taste: Peppery on the palate, with lots of vanilla with some nuts and pitted fruit.
Finish: Long finish with no burn.

Notes

Let this one open up in the glass for full flavor. Perhaps a drop of water would be useful, too. But no more than a drop.

My Score

Notes: ________________________________

____

____

Appearance (1-5) ____ Nose (1-25) ____
Taste (1-25) ____ Finish (1-25) ____
Complexity (1-10) ____ Overall Impression (1-10) ____

SCORE ____

Johnny Drum Green Label

Proof:	80
Age:	NAS
Type:	Straight Bourbon
Style:	Traditional
Mash Bill:	Proprietary
Color:	Straw
Price:	$$

Mike

Nose: Fruit and spice – apricots and pepper. Very light.

Taste: Caramel and apricot with pepper spice and a little hint of oak.

Finish: Long. Starts sweet, but gets peppery and dry with oak tannins.

Susan

Nose: Caramel candy with a tiny bit of vanilla. Strongly reminiscent of Sugar Daddy candy.

Taste: Caramel dominates the palate as well. Not complex but flavorful at 80 proof.

Finish: Smooth with, you guessed, lots of caramel. Dries to a spicy end.

Notes

Caramel lovers will want this on their bar shelves. Enjoy with a chocolate dessert.

My Score

Notes: ______________________________

____

____

Appearance (1-5) ____ Nose (1-25) ____

Taste (1-25) ____ Finish (1-25) ____

Complexity (1-10) ____ Overall Impression (1-10) ____

SCORE ____

Johnny Drum Private Stock

Proof: 101
Age: NAS
Type: Straight Bourbon
Style: Traditional
Mash Bill: Proprietary
Color: Amber
Price: $$

Mike

Nose: Smores. Caramel, marshmallow with a bit of chocolate, a little ripe apple and oak.
Taste: Caramel and milk chocolate with apples and pepper spice. Not hot, nice warmth in the throat.
Finish: Starts sweet and dries out with some oak and pepper spice.

Susan

Nose: Honeycomb, caramel, almonds, dried cherries.
Taste: Drier on the palate than on the nose. Oaky vanilla and nuts.
Finish: Medium long. Smooth. No burn.

Notes

Water brings out more fruit, especially cherries and some chocolate. Very nice with beef entrée or a cigar.

My Score

Notes: ______________________________
____
____

Appearance (1-5) ____ Nose (1-25) ____
Taste (1-25) ____ Finish (1-25) ____
Complexity (1-10) ____ Overall Impression (1-10) ____
SCORE ____

Kentucky Vintage

Proof:	90
Age:	NAS
Type:	Straight
Style:	Small Batch
Mash Bill:	Proprietary
Color:	Dark Straw
Price:	$$

Mike

Nose: Vanilla and caramel with some dates and cherries. Small oak and floral notes.

Taste: Sweet vanilla and berries with some pepper spice.

Finish: Starts sweet and quickly becomes dry with spice and oak. last a long time.

Susan

Nose: Ripe apples and light spices with sweet oak.

Taste: Warm vanilla and walnuts. Not as much fruit as in the nose.

Finish: Long, warm, peppery.

Notes

Batch No. 13-30.

My Score

Notes: ______________________________

____

____

Appearance (1-5) ____ Nose (1-25) ____

Taste (1-25) ____ Finish (1-25) ____

Complexity (1-10) ____ Overall Impression (1-10) ____

SCORE ____

Noah's Mill

Proof: 114.3
Age: NAS
Type: Straight
Style: Small Batch
Mash Bill: Proprietary
Color: Light Amber
Price: $$$

Mike

Nose: Milk chocolate and caramel with some ripe peaches and tobacco. Some oak wood and very pleasant spice.
Taste: Caramel and apricots with oak and pepper spice. Water brings out the fruit and tames the pepper.
Finish: Long and dry. Water amplifies the oak.

Susan

Nose: Honey, caramel, cinnamon and nutmeg.
Taste: Add water to reveal sweet toffee, raisins, dates, and vanilla.
Finish: Long and smooth. No burn here.

Notes

Water brings out all the layers of flavor in this high proof sip. Batch QBC No. 14-61.

My Score

Notes: ______________________________

____

____

Appearance (1-5) ____ Nose (1-25) ____
Taste (1-25) ____ Finish (1-25) ____
Complexity (1-10) ____ Overall Impression (1-10) ____

SCORE ____

Old Bardstown 80

Proof: 80
Age: NAS
Type: Straight Bourbon
Style: Traditional
Mash Bill: Not available
Color: Straw
Price: $

Mike

Nose: Nice nose for 80 proof. Vanilla and honeysuckle with a hint of oak.
Taste: A little thin, but still sweet with caramel, apple fruit, and honey.
Finish: Starts a bit sweet and fades to some oak. Ends quickly.

Susan

Nose: Green apples and a little vanilla. Quite light.
Taste: Roasted, buttered corn. Light oak and a little honey.
Finish: Dries to black pepper and oak.

Notes

One of the first releases from Willett distilled on site. Quite nice for its low proof.

My Score

Notes: ______________________________

____

____

Appearance (1-5) ____ Nose (1-25) ____
Taste (1-25) ____ Finish (1-25) ____
Complexity (1-10) ____ Overall Impression (1-10) ____

SCORE ____

Old Bardstown 90

Proof: 90
Age: NAS
Type: Straight
Style: Traditional
Mash Bill: Not available
Color: Light Amber
Price: $

Mike

Nose: Orange blossoms, caramel, and baking spice.
Taste: Oranges, caramel, cinnamon, and spice.
Finish: Very nice! Starts spicy, but dries out with some sweet oak.

Susan

Nose: Corn, nutmeg, a little honey.
Taste: Caramel corn and sweet nuts. Some orange.
Finish: Warm and smooth with sweet oak.

Notes

One of the first releases from Willett distilled on site. Very attractive price, about $20 per bottle. Willett says they are using a low barrel entry proof.

My Score

Notes: ____________________________________

____

____

Appearance (1-5) ____ Nose (1-25) ____
Taste (1-25) ____ Finish (1-25) ____
Complexity (1-10) ____ Overall Impression (1-10) ____
SCORE ____

Old Bardstown Bottled-in-Bond

Proof: 100
Age: At least 4 years
Type: Straight
Style: Bonded
Mash Bill: Not available
Color: Dark Straw
Price: $

Mike

Nose: Vanilla, nutmeg, allspice. A bit of fruit and oak.
Taste: Nice mouthfeel, almost chewy. With caramel and apples leading into some oak pepper.
Finish: Long and interesting with oak spice. Ends quickly.

Susan

Nose: Caramel, vanilla, dark cherries.
Taste: Smooth sip with caramel, cocoa, some fruit and toasted oak.
Finish: Long, smooth, dry. very pleasant.

Notes

Water makes the bourbon sweeter, but one-dimensional. Made by Willett at its Bardstown distillery.

My Score

Notes: ______________________________

____

____

Appearance (1-5) ____ Nose (1-25) ____
Taste (1-25) ____ Finish (1-25) ____
Complexity (1-10) ____ Overall Impression (1-10) ____

SCORE ____

Old Bardstown Estate Bottled

Proof: 101
Age: NAS
Type: Straight Bourbon
Style: Traditional
Mash Bill: Proprietary
Color: Light Amber
Price: $$

Mike

Nose: Corn, vanilla, caramel with apples and pears and a hint of oak.
Taste: Corn, vanilla, caramel toffee with some apple, pepper spice and fine leather.
Finish: Starts peppery, but gets sweet before drying with some nice oak tannins.

Susan

Nose: Caramel, old leather, sweet oak, orange peel.
Taste: Caramel, toffee, citrusy spice with a touch of honey.
Finish: Sweet at the front and smooths to a long, dry ending.

Notes

Very smooth for the proof. Add a little, but very little, water to bring out more fruit. Limited availability outside of Kentucky.

My Score

Notes: ____________________

____

____

Appearance (1-5) ___ Nose (1-25) ___
Taste (1-25) ___ Finish (1-25) ___
Complexity (1-10) ___ Overall Impression (1-10) ___

SCORE ___

Pure Kentucky Extra Aged

Proof: 107
Age: NAS
Type: Straight
Style: Small Batch
Mash Bill: Proprietary
Color: Light Amber
Price: $$

Mike

Nose: Sweet with caramel and oak, a hint of leather and sweet spices.
Taste: Fruity with berries and apricots. Vanilla, caramel and sweet spice – cinnamon, allspice. Some leather and oak.
Finish: Very nice. Starts sweet and gets spicy and dry with oak.

Susan

Nose: Caramel with perhaps a little leather. Dried fruit emerges after a couple of minutes.
Taste: Vanilla and nuts. Very like a candy bar, but not too sweet. Fruit blossoms with water.
Finish: Long and sweet with a spicy kick at the end.

Notes

Absolutely add a little water (or an ice cube) to bring out all the fruity/spicy character in this bourbon. The high proof can handle it.

My Score

Notes: ____

____

____

Appearance (1-5) ____ Nose (1-25) ____
Taste (1-25) ____ Finish (1-25) ____
Complexity (1-10) ____ Overall Impression (1-10) ____

SCORE ____

Rowan's Creek

Proof: 100.1
Age: NAS
Type: Straight Bourbon
Style: Small Batch
Mash Bill: Proprietary
Color: Straw
Price: $$$

Mike

Nose: Vanilla and bread dough. Not overly complex. Hint of apples and hazelnuts.
Taste: Apricots and caramel with some some pepper spice and pipe tobacco.
Finish: Long and nice, starts sweet with fruit and dries to oak tannins.

Susan

Nose: Ripe apples and pears. Caramel. Fruit dominates, but a little yeasty, too.
Taste: Caramel with a touch of tobacco and new leather. Bright and peppery on the tongue.
Finish: Long and quite smooth for the proof. Dries to sweet oak.

Notes

Very good with a cigar.

My Score

Notes: ____
____
____

Appearance (1-5) ____ Nose (1-25) ____
Taste (1-25) ____ Finish (1-25) ____
Complexity (1-10) ____ Overall Impression (1-10) ____
SCORE ____

Willett Pot Still Reserve

Proof: 94
Age: NAS "aged until fully mature."
Type: Straight Bourbon
Style: Traditional
Mash Bill: Proprietary
Color: Dark Straw
Price: $$$

Mike

Nose: Corn, vanilla, caramel and sweet pipe tobacco.
Taste: Sweet corn, caramel and some pepper spice with a hint of tobacco. A bit chewy. Nice. No burn.
Finish: Starts peppery. Dries to nice oak.

Susan

Nose: Vanilla with some dark cherry and sprinkling of pumpkin pie spices.
Taste: Cherries, vanilla, toasted marshmallow. Rich mouthfeel.
Finish: Sweet to start and lingers to light peppery ending. Not at all hot.

Notes

Probably the most eye-catching bottle on the market. A bonus that the bourbon within is good, too.

My Score

Notes: ____

____

____

Appearance (1-5)	____	Nose (1-25)	____
Taste (1-25)	____	Finish (1-25)	____
Complexity (1-10)	____	Overall Impression (1-10)	____

SCORE ____

Willett Distillery Additional Releases

Name: ______________________________

Distillery: ______________________________

Proof: ____________ **Mash Bill:** ____________
Age: ____________ ____________
Type: ____________ **Color:** ____________
Style: ____________ **Price:** ____________

My Score

Notes: ______________________________

____

____

Appearance (1-5) ____ Nose (1-25) ____
Taste (1-25) ____ Finish (1-25) ____
Complexity (1-10) ____ Overall Impression (1-10) ____
SCORE ____

Name: ______________________________

Distillery: ______________________________

Proof: ____________ **Mash Bill:** ____________
Age: ____________ ____________
Type: ____________ **Color:** ____________
Style: ____________ **Price:** ____________

My Score

Notes: ______________________________

____

____

Appearance (1-5) ____ Nose (1-25) ____
Taste (1-25) ____ Finish (1-25) ____
Complexity (1-10) ____ Overall Impression (1-10) ____
SCORE ____

Other Bourbons

There are hundreds of bourbons being produced by independent distilleries making relatively small amounts of whiskey, and in many cases, other spirits as well. Some of these brands may have a Kentucky address and even be bottled in Kentucky, but the bourbon is sourced from elsewhere, including from MGP (Midwest Grain Products) in Lawrenceburg, Indiana (not to be confused with Lawrenceburg, Kentucky, home to both Four Roses and Wild Turkey.) This facility was formerly known as LDI (Lawrenceburg Distillers Indiana) and was, before that, a Seagrams distillery.

The bourbons are listed alphabetically by name.

11 Wells Minnesota 13 Barrel Aged

Proof: 84
Age: 12 months
Type: Not Straight
Style: Wheated
Mash Bill: 51% corn, 39% wheat, 10% malted barley
Color: Straw
Price: $$$

Mike

Nose: Bread dough and vanilla dominate.
Taste: Vanilla and cereal –Wheat Chex – with a hint of oak.
Finish: Short and dry with oak.

Susan

Nose: Corn and new leather.
Taste: Corn flakes, vanilla, some sweet spice.
Finish: Starts sweet and dries to resinous oak.

Notes

From 11 Wells Spirits Company of St. Paul, Minnesota. Sold in 375ml bottles.

My Score

Notes: ______________________________________
____
____

Appearance (1-5) ____ Nose (1-25) ____
Taste (1-25) ____ Finish (1-25) ____
Complexity (1-10) ____ Overall Impression (1-10) ____
SCORE ____

15 Stars Timeless Reserve

Proof: 103
Age: 13 years
Type: Straight
Style: Small Batch, Blend
Mash Bill: Not available
Color: Dark Straw
Price: $$$$

Mike

Nose: Caramel, apples, pears, oakwood.
Taste: Caramel, pears, with cardamom, spice, oak.
Finish: Long with oak, and cardamom and lingering pears.

Susan

Nose: Caramel, pears, chewable vitamins, light oak.
Taste: Fruity, with white pepper, oak, and vanilla.
Finish: Long and warm.

Notes

Sourced and bottled by 15 Stars, Bardstown, Kentucky. Only 1,800 bottles produced, hence the hefty price tag. There is some 15 year bourbon in the bottle, too.

My Score

Notes: ______________________________

____

____

Appearance (1-5) ____ Nose (1-25) ____
Taste (1-25) ____ Finish (1-25) ____
Complexity (1-10) ____ Overall Impression (1-10) ____
SCORE ____

15 Stars Triple Cask II

Proof: 103
Age: 8 years
Type: Straight
Style: Small Batch, Blend
Mash Bill: Not available
Color: Dark Straw
Price: $$$$

Mike

Nose: Vanilla, raisins, brown sugar, hint of baking spice, and oak.
Taste: Vanilla, brown sugar, raisins, plums, allspice, oak.
Finish: Long with oak and spice that becomes black pepper.

Susan

Nose: Peaches, apple peel, baking spices, caramel.
Taste: Brown sugar, dried apples, some cashews, and baking spices.
Finish: Long and dry.

Notes

Sourced and bottled by 15 Stars, Bardstown, Kentucky. A blend of Kentucky straight bourbons finished in cognac, rum, and apple brandy barrels aged 8 and 16 years.

My Score

Notes: ____________________________________

____

____

Appearance (1-5) ___ Nose (1-25) ___
Taste (1-25) ___ Finish (1-25) ___
Complexity (1-10) ___ Overall Impression (1-10) ___

SCORE ___

18th Street Distillery Bourbon

Proof:	110
Age:	NAS
Type:	Straight
Style:	Barrel Proof
Mash Bill:	Not available
Color:	Straw
Price:	$$

Mike

Nose: Vanilla, banana, oak.
Taste: Vanilla, banana, pepper, and oak.
Finish: Medium long with oak and pepper.

Susan

Nose: Caramel candy, fruit cocktail, very light baking spice.
Taste: Rich caramel cinnamon red hots, little fruit.
Finish: Medium long and peppery.

Notes

18th Street Distillery, Hammond, Indiana.

My Score

Notes: ____

____

____

Appearance (1-5)	____	Nose (1-25)	____
Taste (1-25)	____	Finish (1-25)	____
Complexity (1-10)	____	Overall Impression (1-10)	____

SCORE ____

A.D. Laws Four Grain Bourbon

Proof: 95
Age: At least 3 years
Type: Straight
Style: Four Grain
Mash Bill: 60% corn, 20% wheat, 10% rye, 10% malted barley
Color: Amber
Price: $$$

Mike

Nose: Caramel with a hint of dark chocolate.
Taste: Caramel and chocolate with a little apricot and baking spice.
Finish: Long and dry with oak and chocolate.

Susan

Nose: Stewed prunes, nuts, and oak with a little chocolate.
Taste: Cereal medley with a sprinkling of brown sugar and light apricot note.
Finish: Dry and spicy.

Notes

Very rich for its age. Heirloom grains are used. Laws Whiskey House, Denver, Colorado, https://lawswhiskeyhouse.com.

My Score

Notes: ______________________________

____

____

Appearance (1-5) ____ Nose (1-25) ____
Taste (1-25) ____ Finish (1-25) ____
Complexity (1-10) ____ Overall Impression (1-10) ____

SCORE ____

A.D. Laws Four Grain Bourbon Bottled-in-Bond

Proof: 100
Age: At least 4 years
Type: Straight
Style: Bottled-in-Bond
Mash Bill: 60% corn, 20% wheat, 10% rye, 10% malted barley
Color: Deep Amber
Price: $$$

Mike

Nose: Caramel with a hint of chocolate and pecans.
Taste: Chocolate with baking spices and nuts – pecans or hazelnuts. Keeps switching as I drink.
Finish: Long and dry with dark chocolate and nuts.

Susan

Nose: Tobacco, chocolate-covered berries, light caramel.
Taste: Toffee, new leather, cinnamon, nutmeg, and dark cherries.
Finish: Dries to a peppery, herbal ending.

Notes

Rich mouthfeel. Uses heirloom grains. Laws Whiskey House, Denver, Colorado, https://lawswhiskeyhouse.com.

My Score

Notes: ______________________________
____
____

Appearance (1-5) ____ Nose (1-25) ____
Taste (1-25) ____ Finish (1-25) ____
Complexity (1-10) ____ Overall Impression (1-10) ____

SCORE ____

Amador Double Barrel

Proof: 86.8
Age: NAS
Type: Not Straight
Style: Special Finish
Mash Bill: Not available
Color: Straw
Price: $$

Mike

Nose: Raisins and dates. Vanilla and sweet spice.
Taste: Thin and watery with some fruit and vanilla.
Finish: Dry and oaky.

Susan

Nose: Very light. Corn and some cherries.
Taste: Fruit note, probably from wine barrel.
Finish: Short and hot.

Notes

Sourced from undisclosed Kentucky distillery and finished in Napa Valley wine barrels by Amador Whiskey Co., St. Helena, California.

My Score

Notes: ______________________________

____

____

Appearance (1-5) ____ Nose (1-25) ____
Taste (1-25) ____ Finish (1-25) ____
Complexity (1-10) ____ Overall Impression (1-10) ____

SCORE ____

Andy's Old No. 5

Proof: 90
Age: NAS
Type: Not Straight
Style: Small Batch
Mash Bill: Not available
Color: Very Light Straw
Price: $$$

Mike

Nose: Bread dough and corn, a hint of vanilla.
Taste: Corn and vanilla with a hint of bread dough.
Finish: Dry with oak and pepper.

Susan

Nose: Corn and oak.
Taste: Dry cereal with a dash of cinnamon.
Finish: Oaky with a dash of pepper.

Notes

From Indian Creek Distillery, Staley Mill Farm and Distillery of New Carlisle, Ohio. This was Bottle 50 from Batch Number 2.

My Score

Notes: ____________________

____

____

Appearance (1-5) ____ Nose (1-25) ____
Taste (1-25) ____ Finish (1-25) ____
Complexity (1-10) ____ Overall Impression (1-10) ____
SCORE ____

Baby Jane

Proof: 91
Age: NAS
Type: Straight
Style: Blend of Straight Bourbons
Mash Bill: Not available
Color: Straw
Price: $$

Mike

Nose: Vanilla, dried fruit, baking spice, and oak.
Taste: Vanilla, dates, apricots, cinnamon, nutmeg, oak.
Finish: Medium long with oak and spice.

Susan

Nose: Caramel, vanilla, apple, a little cardamom.
Taste: Vanilla, savory herbs, some peach.
Finish: Medium long, warm, ending in soft oak.

Notes

Widow Jane Distillery, Brooklyn, New York. Uses heirloom corn.

My Score

Notes: ______________________________

____

____

Appearance (1-5) ____ Nose (1-25) ____
Taste (1-25) ____ Finish (1-25) ____
Complexity (1-10) ____ Overall Impression (1-10) ____

SCORE ____

Backbone Bourbon

Proof:	115.4
Age:	NAS
Type:	Straight
Style:	Uncut, Barrel Strength, Small Batch
Mash Bill:	Not available
Color:	Dark Straw
Price:	$$$

Mike

Nose: Sweet caramel, ginger and nutmeg. Water softens the spices.

Taste: Sweet and spice with molasses and spice. Water makes it sweeter.

Finish: Long, sweet, spicy. Water makes it peppery.

Susan

Nose: Crème brulee, ripe apple/pear. Faint note of cocoa.

Taste: Sweet vanilla, sweet spices, baked apples.

Finish: Warm with black pepper and oak at the end.

Notes

Bottled by Bardstown Backbone Bourbon Co. This was small batch #10. Distilled at MGP in Lawrenceburg, Indiana.

My Score

Notes: ______________________________

____

____

Appearance (1-5) ____ Nose (1-25) ____

Taste (1-25) ____ Finish (1-25) ____

Complexity (1-10) ____ Overall Impression (1-10) ____

SCORE ____

Balcones Texas Blue Corn Bourbon

Proof: 129.8
Age: 24 months
Type: Straight
Style: Non-Chill Filtered, Barrel Strength
Mash Bill: 100% blue corn
Color: Amber
Price: $$

Mike

Nose: Corn, vanilla, and some ginger. Water opens it up with more corn.
Taste: Caramel corn and baking spices. Water brings out tannins.
Finish: Long and sweet with a hint of pepper.

Susan

Nose: Sweet caramel corn, milk chocolate, berry fruit.
Taste: Licorice, baking spices, apples.
Finish: Long, sweet, with apricot notes.

Notes

Made by Balcones Distilling, Waco, Texas. Add one ice cube to open it up and reveal fruit. Any more water or ice flattens it and removes complexity.

My Score

Notes: ______________________________

____

____

Appearance (1-5) ____ Nose (1-25) ____
Taste (1-25) ____ Finish (1-25) ____
Complexity (1-10) ____ Overall Impression (1-10) ____
SCORE ____

Balcones Texas Pot Still Bourbon

Proof: 92
Age: 24 months
Type: Straight
Style: Pot Distilled; Non-Chill Filtered
Mash Bill: Undisclosed amounts of blue corn, Texas wheat, Texas rye, and malted barley
Color: Dark Straw
Price: $$

Mike

Nose: Fruity with apricots and pears, vanilla, hints of oak and milk chocolate.
Taste: Fruity with apricots, caramel, baking spices, and chocolate. Just a hint of oak toward the finish.
Finish: Medium long with oak and chocolate.

Susan

Nose: Grain forward followed by ripe apricots and burnt caramel.
Taste: Rich caramel with apricots flavor in equal measure along with some raisin notes.
Finish: Medium long and very soft, drying to oak.

Notes

Whiskey tasted was bottled 28 March, 2019. Balcones Distilling, Waco, Texas, https://balconesdistilling.com.

My Score

Notes: ______________________________________

____

____

Appearance (1-5) ____ Nose (1-25) ____
Taste (1-25) ____ Finish (1-25) ____
Complexity (1-10) ____ Overall Impression (1-10) ____
SCORE ____

Banker's Club

Proof: 80
Age: NAS
Type: Straight
Style: Traditional
Mash Bill: Not available
Color: Light Straw
Price: $

Mike

Nose: Vanilla and sweet corn. Very light with just a hint of oak.
Taste: Sweet spice and vanilla. Not complex. Some corn and oak.
Finish: Moderately long. A bit spicy with oak tannins.

Susan

Nose: Minty, light, with a whiff of spice.
Taste: Light oak with some spice and a little vanilla.
Finish: Short and oddly hot for 80 proof.

Notes

Not bad for the price. Use it to make cocktails for a crowd. Distributed by Regent Distillers Products Company, Scobeyville, New Jersey. Made in Kentucky.

My Score

Notes: ______________________________________

____

____

Appearance (1-5) ____ Nose (1-25) ____
Taste (1-25) ____ Finish (1-25) ____
Complexity (1-10) ____ Overall Impression (1-10) ____

SCORE ____

Bare Knuckle Straight Bourbon

Proof: 90
Age: 2 years
Type: Straight
Style: Wheated
Mash Bill: 70% corn, 20% wheat, 10% malted barley
Color: Dark Straw
Price: $$

Mike

Nose: Cornbread dough with some vanilla and a hint of apples.
Taste: Corn, vanilla, a little baking spice and a hint of apples.
Finish: Medium long with spices and a little oak.

Susan

Nose: Roasted corn, light custard, and some ripe apple.
Taste: Corn and vanilla with some hard candy and oak.
Finish: Dry and oaky.

Notes

Other expressions include Distiller's Reserve Bourbon and High Rye Bourbon. KO Distillling, Manassas, Virginia.

My Score

Notes: ____

____

____

Appearance (1-5) ____ Nose (1-25) ____
Taste (1-25) ____ Finish (1-25) ____
Complexity (1-10) ____ Overall Impression (1-10) ____
SCORE ____

Barrell Armida

Proof: 107.6
Age: NAS
Type: Straight
Style: Barrel Proof, Blend, Finished
Mash Bill: Not available
Color: Dark Straw
Price: $$$

Mike

Nose: Vanilla, baking spices, pears, and berries.
Taste: Vanilla, some anise, a hint of tobacco, and nuts.
Finish: Long, drying to oaky tannins.

Susan

Nose: Vanilla, some almond, baking spices, and pears.
Taste: Still vanilla, a little herbal, with licorice, pipe tobacco, and nuts.
Finish: Long and dry, almost mineral.

Notes

Barrell Craft Spirits, Louisville, Kentucky. One of Barrell's regular releases. Bourbons from Kentucky and Indiana are finished in pear brandy, rum, and Sicilian amaro casks. Visit the website to see all releases.

My Score

Notes: ____

____

____

Appearance (1-5)	____	Nose (1-25)	____
Taste (1-25)	____	Finish (1-25)	____
Complexity (1-10)	____	Overall Impression (1-10)	____

SCORE ____

Barrell Bourbon

Proof:	124.7
Age:	8 years, 3 months
Type:	Straight
Style:	Small Batch, Barrel Proof
Mash Bill:	70% corn, 26% rye, 4% malted barley
Color:	Dark Straw
Price:	$$$

Mike

Nose: Maple smoke and apples with a hint of vanilla.
Taste: Sweet caramel with smoky wood and apples. Hint of pepper spice.
Finish: Dry and smoky with oak.

Susan

Nose: Caramel corn. Apples.
Taste: Very sweet entry – candy corn, maple syrup, cinnamon.
Finish: Cinnamon grows at the end.

Notes

Distilled in Tennessee (George Dickel?). Bottled by Barrel Craft Spirits, Crestwood, Kentucky. This was batch 005. Tasted a lot like juice from George Dickel Distillery. Each batch sourced differently. A very pleasant sip. Water amplifies the apples. Barrel announced plans in 2017 to built its own distillery in Louisville.

My Score

Notes: ______________________________

____

____

Appearance (1-5) ____ Nose (1-25) ____
Taste (1-25) ____ Finish (1-25) ____
Complexity (1-10) ____ Overall Impression (1-10) ____

SCORE ____

Barrell Bourbon Cask Strength Batch 15

Proof:	107.6
Age:	9 years, 6 months (youngest)
Type:	Blend
Style:	Barrel Proof, Extra Aged
Mash Bill:	Batched (See notes)
Color:	Pale Straw
Price:	$$$

Mike

Nose: Vanilla and ripe pears with a hint of oak.
Taste: Vanilla and pears with some baking spices.
Finish: Medium long and dry with some oak and spice.

Susan

Nose: Caramel, some sweet spice, stewed fruit.
Taste: Caramel and cinnamon with light sweet mint and pears.
Finish: Long and peppery drying to oak.

Notes

Brand specializes in sources whiskeys to create unique batches, no two of which will be alike. This batch from 2018 contains 9 year and 6 month, 10 and 11 year bourbons from Tennessee and Kentucky.

My Score

Notes: ______________________________

____

____

Appearance (1-5) ____ Nose (1-25) ____
Taste (1-25) ____ Finish (1-25) ____
Complexity (1-10) ____ Overall Impression (1-10) ____

SCORE ____

Barrell Bourbon Cask Strength Batch 16

Proof: 105.8
Age: 9 years, 9 months (youngest)
Type: Blend
Style: Barrel Proof, Finished
Mash Bill: Batched (See notes)
Color: Pale Straw
Price: $$$

Mike

Nose: Very light nose. Vanilla and a hint of oak.
Taste: Light fruit – berries and dates – vanilla and lots of oak tannins.
Finish: Surprisingly short with oak tannins.

Susan

Nose: Vanilla and a bit of sweet apple.
Taste: Cherries, vanilla, dash of cocoa.
Finish: Dry and peppery.

Notes

Water released lots of fruit and tamed the oak. This batch from 2019 contains 9 year and 9 month, 11 year, and 15 year bourbons from Kentucky, Indiana, and Tennessee.

My Score

Notes: ____________________
____
____

Appearance (1-5) ____ Nose (1-25) ____
Taste (1-25) ____ Finish (1-25) ____
Complexity (1-10) ____ Overall Impression (1-10) ____

SCORE ____

Barrell Bourbon Cask Strength Batch 17

Proof: 112.5
Age: 10 years, 4 months (youngest)
Type: Blend
Style: Barrel Proof, Extra Aged
Mash Bill: Batched (See notes)
Color: Straw
Price: $$$

Mike

Nose: Vanilla and berries with a hint of oak.
Taste: Vanilla and berries with baking spices – cinnamon and nutmeg – with some oak.
Finish: Long – starts sweet but dries out with oak and spice.

Susan

Nose: Caramel, orchard fruit, some spice.
Taste: Cocoa, caramel, and oak.
Finish: Long and dry with oak tannins.

Notes

This batch from 2019 contains 10 year and 4 month, 15 year and 15 year bourbons from Indiana, Tennessee, and Kentucky.

My Score

Notes: ______________________________

____

____

Appearance (1-5) ____ Nose (1-25) ____
Taste (1-25) ____ Finish (1-25) ____
Complexity (1-10) ____ Overall Impression (1-10) ____

SCORE ____

Battlefield Bourbon

Proof: 91.2
Age: NAS
Type: Not Straight
Style: Small Batch
Mash Bill: Not available
Color: Very Light Straw
Price: $$$

Mike

Nose: Corn and vanilla with a hint of pecans.
Taste: Corn, vanilla, honey with some citrus notes.
Finish: Short and sweet with a hint of oak.

Susan

Nose: Sweet corn, nuts, nutmeg, orange zest.
Taste: Sweet spices, citrus, vanilla.
Finish: Sweet, drying to oak and cornmeal.

Notes

Speakeasy Spirits LLC, Nashville. Distilled, aged and bottled in Tennessee. This was bottle #882 of 1863. Proceeds help preserve and restore Civil War battlefield sites.

My Score

Notes: ____________________

____

____

Appearance (1-5) ____ Nose (1-25) ____
Taste (1-25) ____ Finish (1-25) ____
Complexity (1-10) ____ Overall Impression (1-10) ____

SCORE ____

Beanball Bourbon

Proof: 90
Age: NAS
Type: Not Straight
Style: Small Batch
Mash Bill: Not available
Color: Light Straw
Price: $$$

Mike

Nose: Corn bread and vanilla/honey.
Taste: Peaches and vanilla with a hint of spice.
Finish: Not much. A little sweet and gone.

Susan

Nose: Sweet oak and Bit O'Honey.
Taste: Ripe peaches and some vanilla.
Finish: Short and light.

Notes

Bottle number 6-8-15. Made in Indiana at MGP and bottled by Cooperstown Distillery, Cooperstown, New York.

My Score

Notes: ______________________________

____

____

Appearance (1-5) ___ Nose (1-25) ___
Taste (1-25) ___ Finish (1-25) ___
Complexity (1-10) ___ Overall Impression (1-10) ___

SCORE ___

Beer Barrel Bourbon

Proof: 80
Age: NAS
Type: Not Straight
Style: Special Finish
Mash Bill: Not available
Color: Straw
Price: $$$

Mike

Nose: Very malty. Not complex. Had a hard time getting anything else. A little vanilla

Taste: Very light. Some malt and chocolate notes with some caramel and corn.

Finish: Very short and light A little sweet with a hint of oak.

Susan

Nose: Light and malty. Some oak and faint vanilla. Not complex.

Taste: Caramel and sweet spices.

Finish: Short and peppery. Oddly hot for 80 proof.

Notes

New Holland Artisan Spirits, Holland, Michigan. New Holland is known for its craft beers. This bourbon is sourced from Kentucky and aged for an additional three months in used beer barrels. Might make an interesting Manhattan.

My Score

Notes: ____________________

____

____

Appearance (1-5) ____ Nose (1-25) ____

Taste (1-25) ____ Finish (1-25) ____

Complexity (1-10) ____ Overall Impression (1-10) ____

SCORE ____

Belle Meade

Proof:	90.4
Age:	5 ½ - 7 ½ years
Type:	Straight
Style:	Small Batch
Mash Bill:	30% rye
Color:	Light Amber
Price:	$$

Mike

Nose: Corn and vanilla with a hint of caramel and oak.
Taste: Corn, nutmeg, clove, with some vanilla oak.
Finish: Dry with oak tannins and a bit astringent.

Susan

Nose: Maple syrup, cinnamon, and vanilla.
Taste: Caramel corn, very dry stone fruit and new oak.
Finish: Short and dry.

Notes

Bottled by Nelson's Green Brier Distillery, Nashville, Tennessee. Sourced from MGP in Lawrenceburg, Indiana. Belle Meade Bourbon distilled at Green Brier to be released in 2015.

My Score

Notes: ____________________________________

____

____

Appearance (1-5) ____ Nose (1-25) ____
Taste (1-25) ____ Finish (1-25) ____
Complexity (1-10) ____ Overall Impression (1-10) ____

SCORE ____

Belle Meade Madeira Finish

Proof: 90.4
Age: NAS
Type: Finished
Style: Small Batch, Special Finish
Mash Bill: Not available
Color: Light Amber
Price: $$$

Mike

Nose: Raisins and cherries with caramel and baking spices.
Taste: Fruity with raisins and dates, a bit of caramel, baking spice and oak.
Finish: Long and dry. Starts fruity, but turns to nice oak.

Susan

Nose: Sweet wood smoke, fruit, light caramel.
Taste: Fruity caramel. Madeira note accounts for sweetness.
Finish: Dry and peppery.

Notes

From Nelson's Green Brier Distillery, Nashville, Tennessee. This was Batch 1, Bottle Number 1432.

My Score

Notes: ____

____

____

Appearance (1-5) ____ Nose (1-25) ____
Taste (1-25) ____ Finish (1-25) ____
Complexity (1-10) ____ Overall Impression (1-10) ____
SCORE ____

Belle Meade Sherry Cask

Proof:	90.4
Age:	9 years
Type:	Finished
Style:	Small Batch, Special Finish
Mash Bill:	Not available
Color:	Light Amber
Price:	$$$

Mike

Nose: Berries and chocolate with a hint of oak.
Taste: Berries, caramel, vanilla. Hint of baking spice.
Finish: Fruity start which dries out to oak and spice.

Susan

Nose: Sweetly nutty, with vanilla and caramel.
Taste: Berries, caramel, almonds, sweet spice.
Finish: Dry with figs at the end.

Notes

From Nelson's Green Brier Distillery, Nashville, Tennessee. Finished in Oloroso sherry casks. Small Batch Number 1 and Bottle 492.

My Score

Notes: ____________________________________

____

____

Appearance (1-5)	____	Nose (1-25)	____
Taste (1-25)	____	Finish (1-25)	____
Complexity (1-10)	____	Overall Impression (1-10)	____

SCORE ____

Ben Holladay Missouri Straight Bourbon

Proof:	100
Age:	6 years
Type:	Straight
Style:	Bonded
Mash Bill:	Not available
Color:	Straw
Price:	$$$

Mike

Nose: Vanilla, apples, baking spice, oak.

Taste: Vanilla, apples, raspberry, cinnamon, nutmeg, oak.

Finish: Medium long with oak, spice, and lingering fruit.

Susan

Nose: Vanilla, ripe melon, passionfruit, some allspice, and new leather.

Taste: Vanilla and spice. Not as much fruit as on the nose, but there. Leather still present.

Finish: Long and dry with a touch of fruit.

Notes

Revived brand from the historic McCormick Distillery, Weston, Missouri. Holladay was also a carriage maker, hence the label image.

My Score

Notes: ____________________________________

____

____

Appearance (1-5) ____ Nose (1-25) ____

Taste (1-25) ____ Finish (1-25) ____

Complexity (1-10) ____ Overall Impression (1-10) ____

SCORE ____

Benjamin Prichard's Double Barrel

Proof: 90
Age: 9 years
Type: Not Straight
Style: Special Finish
Mash Bill: Not available
Color: Light Amber
Price: $$

Mike

Nose: Vanilla and toffee with a hint of pear.
Taste: Pears and caramel with some pepper and oak.
Finish: Starts fruity and dries to oak.

Susan

Nose: Caramel, light pear, sweet nuts.
Taste: Caramel and nuts with a little pear.
Finish: Warm and smooth with oak note at the end.

Notes

Prichard's Distillery, Kelso, Tennessee. Uses copper pot stills. Aged for several years in one barrel and then for a few more in another new, charred oak barrel.

My Score

Notes: ____

____

____

Appearance (1-5) ____ Nose (1-25) ____
Taste (1-25) ____ Finish (1-25) ____
Complexity (1-10) ____ Overall Impression (1-10) ____
SCORE ____

Berkshire Bourbon

Proof: 86
Age: NAS
Type: Not Straight
Style: Pot Stilled, Small Batch, Craft
Mash Bill: Not available
Color: Light Straw
Price: $$$

Mike

Nose: A little butter toffee and leather.
Taste: A little cherry/berry fruit and vanilla with sweet spices. A bit creamy.
Finish: Starts sweet but gets sine oak toward the end. Fairly short.

Susan

Nose: Very light. Grain – rye as well as complex. Faint tobacco.
Taste: Light oak with a little corn sweetness some and sweet spices.
Finish: Medium. Dries to black pepper.

Notes

Berkshire Mountain Distillers, Sheffield, Massachusetts. Would be interesting to see what more age could do for this bourbon. Grains are locally sourced.

My Score

Notes: ______________________________

____

____

Appearance (1-5) ____ Nose (1-25) ____
Taste (1-25) ____ Finish (1-25) ____
Complexity (1-10) ____ Overall Impression (1-10) ____
SCORE ____

Bib & Tucker

Proof: 92
Age: 6 years
Type: Straight
Style: Small Batch
Mash Bill: Not available
Color: Light Amber
Price: $$$

Mike

Nose: Peaches and vanilla with some cinnamon oak and leather. Nuttiness.
Taste: Caramel and apricots. Some allspice. Oak, too.
Finish: Nice finish. Medium long with caramel and oak. Dries out nicely.

Susan

Nose: Corn and rye. Warm cereal. Some spice.
Taste: Caramel, vanilla, nutmeg and cinnamon.
Finish: Long and sweet, ending in sweet oak.

Notes

Bottled by 35 Maple Street Spirits, Crestwood, Kentucky. Very handsome, flask-style bottle.

My Score

Notes: ____

____

____

Appearance (1-5) ____ Nose (1-25) ____
Taste (1-25) ____ Finish (1-25) ____
Complexity (1-10) ____ Overall Impression (1-10) ____
SCORE ____

Big Ass Bourbon Barrel Strength

Proof: 119.5
Age: 4 months
Type: Not Straight
Style: Barrel Strength
Mash Bill: Not available
Color: Very Light Straw
Price: $$$

Mike

Nose: Young. Bread dough and vanilla.
Taste: Corn and vanilla. Very light.
Finish: Pepper and little else.

Susan

Nose: Oak and alcohol. Whiff of vanilla.
Taste: Vanilla. Almost flavorless.
Finish: Oaky and hot.

Notes

Probably an MGP product. Owned by Diversa Brands, Louisville, Kentucky.

My Score

Notes: ____

____

____

Appearance (1-5) ____ Nose (1-25) ____
Taste (1-25) ____ Finish (1-25) ____
Complexity (1-10) ____ Overall Impression (1-10) ____
SCORE ____

Big Bottom Small Batch

Proof: 91
Age: NAS
Type: Straight
Style: Small Batch, Craft
Mash Bill: 36% rye
Color: Dark Straw
Price: $

Mike

Nose: Vanilla with corn and sweet spice – nutmeg and cinnamon.
Taste: Sweet vanilla with some cherries. Nutmeg and cinnamon here, too.
Finish: Long and spicy with oak tannins.

Susan

Nose: Caramel, apple, nutmeg. Very appealing.
Taste: Vanilla and fruit. Almost a sundae.
Finish: Spicy oak and sweet vanilla.

Notes

Made by Big Bottom Distilling, Hillsboro Oregon. Distiller Ted Pappas also makes bourbons finished in port, zinfandel, and cabernet barrels, as well as a 111 proof bourbon. Named for an Oregon nature preserve. Certainly worth seeking out.

My Score

Notes: ____________________________________

____

____

Appearance (1-5) ____ Nose (1-25) ____
Taste (1-25) ____ Finish (1-25) ____
Complexity (1-10) ____ Overall Impression (1-10) ____

SCORE ____

Big House

Proof:	90
Age:	6 years
Type:	Straight Bourbon
Style:	Small Batch
Mash Bill:	60% corn, 35% rye, 5% malted barley
Color:	Dark Straw
Price:	$

Mike

Nose: Caramel and citrus fruit with hint of honeysuckle. Some sweet spice – allspice – in the background.

Taste: A bit of corn and vanilla with some citrus fruit and oak.

Finish: Starts sweet with caramel, but dries with oak tannins and spice.

Susan

Nose: Vanilla, light caramel, sweet licorice and a little new leather.

Taste: Crème brulee, citrus, and faint nuttiness. Creamy mouthfeel.

Finish: Weakest aspect. Rather short and hot.

Notes

Underdog Wine and Spirits, Livermore, California. Sourced from MGP in Indiana. Limited distribution.

My Score

Notes: ______________________________

____

____

Appearance (1-5) ____ Nose (1-25) ____

Taste (1-25) ____ Finish (1-25) ____

Complexity (1-10) ____ Overall Impression (1-10) ____

SCORE ____

Bird Dog

Proof: 80
Age: NAS
Type: Not Straight
Style: Traditional
Mash Bill: Not available
Color: Light Straw
Price: $$

Mike

Nose: Very light – vanilla and honey.
Taste: Very simple. A little vanilla and corn. Hint of pepper spice.
Finish: Short and sweet with vanilla.

Susan

Nose: Light. Corn and alcohol. Vanilla.
Taste: Corn and vanilla. That's it. Hot for 80 proof.
Finish: Short. Heat subsides at the end.

Notes

Three Springs Bottling Company, Bowling Green, Kentucky. Says "Kentucky" on the label, but no distiller identified.

My Score

Notes: ____________________________________

____

____

Appearance (1-5) ____ Nose (1-25) ____
Taste (1-25) ____ Finish (1-25) ____
Complexity (1-10) ____ Overall Impression (1-10) ____
SCORE ____

Bird Dog Small Batch

Proof: 86
Age: NAS
Type: Not Straight
Style: Small Batch
Mash Bill: Not available
Color: Straw
Price: $$

Mike

Nose: Corn and honeysuckle with hint of vanilla.
Taste: Corn and vanilla with pepper spice.
Finish: Short, dry and peppery.

Susan

Nose: Vanilla, pitted fruit, flowers.
Taste: Sweet buttered corn, apples, faint spice.
Finish: Medium long, drying to oak.

Notes

Western Spirits and 3 Springs Bottling Co., Bowling Green, Kentucky.

My Score

Notes: ____________________

____

____

Appearance (1-5) ____ Nose (1-25) ____
Taste (1-25) ____ Finish (1-25) ____
Complexity (1-10) ____ Overall Impression (1-10) ____

SCORE ____

Black Maple Hill Oregon Straight Bourbon Whiskey

Proof: 95
Age: NAS
Type: Straight
Style: Small Batch
Mash Bill: Not available
Color: Pale Straw
Price: $$$

Mike

Nose: Hint of caramel and corn with a little anise and apricot.
Taste: Corn, vanilla, and anise with a little apricot and pepper.
Finish: Medium long with pepper and oak.

Susan

Nose: Caramel corn, some brown sugar and dried fruit.
Taste: Vanilla, apple, caramel, and roasted corn.
Finish: Light pepper and dry oak.

Notes

Limited edition. Stein Distillery, Joseph, Oregon.

My Score

Notes: ________________________________

____

____

Appearance (1-5) ____ Nose (1-25) ____
Taste (1-25) ____ Finish (1-25) ____
Complexity (1-10) ____ Overall Impression (1-10) ____

SCORE ____

Black Ridge Small Batch

Proof: 90
Age: NAS
Type: Straight
Style: Small Batch
Mash Bill: Not available
Color: Dark Straw
Price: $$$

Mike

Nose: Ripe fruit, leather, caramel.
Taste: Ripe apples, pears and caramel with baking spice and tobacco.
Finish: Starts fruity, but gets some nice oak and spice as it lingers. Very nice.

Susan

Nose: Sweet corn, light cinnamon, pitted fruit.
Taste: Vanilla, caramel, wood smoke.
Finish: Sweet, drying to oak and smoke.

Notes

Brand from Clear Springs Distilling Company, Louisville, Kentucky, which is owned by Sazarac.

My Score

Notes: ____________________

____

____

Appearance (1-5) ____ Nose (1-25) ____
Taste (1-25) ____ Finish (1-25) ____
Complexity (1-10) ____ Overall Impression (1-10) ____

SCORE ____

Black Saddle

Proof: 90
Age: 12 years
Type: Does not say "Straight" on label
Style: Traditional
Mash Bill: Not available
Color: Straw
Price: $$$

Mike

Nose: Caramel and apple with a hint of nutmeg and oak.
Taste: Caramel and ripe apples with some pepper spice.
Finish: Long and dry with oak and pepper.

Susan

Nose: Appealing vanilla character.
Taste: Cherry and a little wood smoke. Caramel, some pitted fruit and nuts.
Finish: Dries to pleasant oak spiciness.

Notes

From Frank-Lin Distillers Products Ltd. Fairfield, California. Distributors of several Kentucky-sourced bourbons of indeterminate origin.

My Score

Notes: ______________________________

____

____

Appearance (1-5) ____ Nose (1-25) ____
Taste (1-25) ____ Finish (1-25) ____
Complexity (1-10) ____ Overall Impression (1-10) ____

SCORE ____

Blaum Bros. Straight Bourbon

Proof: 100
Age: NAS
Type: Straight
Style: Non-Chill Filtered
Mash Bill: Not available
Color: Dark Straw
Price: $$

Mike

Nose: Corn and vanilla with leather, tobacco, and oak.
Taste: Corn, vanilla, a little citrus, white pepper and oak.
Finish: Long with oak and pepper.

Susan

Nose: Caramel corn, some brown sugar and dried fruit
Taste: Vanilla, apple, caramel, and roasted corn.
Finish: Light pepper and dry oak.

Notes

In case you were wondering, there are real brothers, who are named Matt and Mike. They started the distillery and act as distillers and blenders. Blaum Bros. Distilling Company Galena, Illinois, https://www.blaumbros.com.

My Score

Notes: ____________________

____

____

Appearance (1-5) ____ Nose (1-25) ____
Taste (1-25) ____ Finish (1-25) ____
Complexity (1-10) ____ Overall Impression (1-10) ____

SCORE ____

Blind Tiger

Proof: 90
Age: NAS
Type: Not Straight
Style: Small Batch
Mash Bill: Not available
Color: Dark Straw
Price: $$$

Mike

Nose: Corn and vanilla. Hint of fruit.
Taste: Corn and vanilla, Very young. A bit thin.
Finish: Sweet corn and vanilla. Very sweet and long.

Susan

Nose: Corn predominates. Whiff of vanilla and some underlying sweetness.
Taste: Corn dominates the palate. A little candy corn sweetness.
Finish: Vanilla shows up here and lingers after a sprinkling of pepper. Very nice finish.

Notes

Chicago Distilling Company, Chicago, Illinois. Quite nice for a young bourbon. We hope future batches will be allowed to age longer. This was Batch Number 1.

My Score

Notes: ______________________________

____

____

Appearance (1-5) ____ Nose (1-25) ____
Taste (1-25) ____ Finish (1-25) ____
Complexity (1-10) ____ Overall Impression (1-10) ____

SCORE ____

Bluegrass Blue Corn Single Barrel Bourbon

Proof: 127.4
Age: 4 years
Type: Straight
Style: Single Barrel, Cask Strength
Mash Bill: 75% blue corn, 21% rye, 4% malted barley
Color: Light Amber
Price: $$$

Mike

Nose: Lots of caramel with a hint of dark chocolate, prunes, dates, and a little oak.
Taste: Caramel, dried apricots, cinnamon, nutmeg, and oak.
Finish: Long with oak and cinnamon.

Susan

Nose: Caramel covered vanilla wafer with some baking spice and dark fruit.
Taste: Fruit forward with apple and pears, followed by caramel and spice.
Finish: Long and warm with a dash of cinnamon red hots.

Notes

Water brings out the dark chocolate and reveals caramel apple. Bluegrass Distillers, Lexington, Kentucky, https://www.bluegrassdistillers.com.

My Score

Notes: ____

____

____

Appearance (1-5) ____ Nose (1-25) ____
Taste (1-25) ____ Finish (1-25) ____
Complexity (1-10) ____ Overall Impression (1-10) ____

SCORE ____

Bluegrass Distillers Bourbon

Proof:	109
Age:	NAS, but less than 2 years
Type:	Not Straight
Style:	Small Batch
Mash Bill:	80% corn, 20% rye
Color:	Dark Straw
Price:	$$

Mike

Nose: Bread dough and vanilla. Very light for the proof.

Taste: Bread dough, vanilla and a hint of fruit and oak.

Finish: Oaky and astringent.

Susan

Nose: Oak. Like sniffing an old log.

Taste: Oak and some corn and pepper.

Finish: Short and woody, but a sweet note a emerges at the end.

Notes

Bluegrass Distillers. Lexington, Kentucky. Aged in 25-gallon barrels with #4 char. Planned release of a straight, 2-year-old bourbon in 2017.

My Score

Notes: ____________________________________

____

____

Appearance (1-5) ___ Nose (1-25) ___

Taste (1-25) ___ Finish (1-25) ___

Complexity (1-10) ___ Overall Impression (1-10) ___

SCORE ___

Bluegrass High Rye Single Barrel Bourbon

Proof: 113.2
Age: 3 years
Type: Straight
Style: Single Barrel, Cask Strength, Pot Distilled
Mash Bill: 75% corn, 21% rye, 4% malted barley
Color: Light Straw
Price: $$$

Mike

Nose: Lots of fruit – plums and bananas – with vanilla and oak.
Taste: Vanilla and plums with allspice, white pepper, oak.
Finish: Long with oak and spice.

Susan

Nose: Multigrain bread dough with a little carraway seed, banana, and brown sugar.
Taste: Caramel and vanilla with some black currants and baking spice.
Finish: Quick.

Notes

Water brings out candied fruit and softens the grains. The distillery also makes a 90 proof expression. Barrel 16, February 2018.

My Score

Notes: ______________________________

____

____

Appearance (1-5) ____ Nose (1-25) ____
Taste (1-25) ____ Finish (1-25) ____
Complexity (1-10) ____ Overall Impression (1-10) ____

SCORE ____

Bluegrass Toasted Oak Bourbon

Proof:	100
Age:	At least 4 years
Type:	Straight
Style:	Secondary Barrel, Pot Distilled
Mash Bill:	75% corn, 21% rye, 4% malted barley
Color:	Straw
Price:	$$$

Mike

Nose: Rich caramel and chocolate with hints of fruit and oak.

Taste: Caramel apples, white pepper, and oak with a hint of cherries.

Finish: Long with oak and pepper and lingering caramel.

Susan

Nose: Toffee, sweet grains, and whiffs of cherries and chocolate.

Taste: Caramel predominates, along with oak and sweet spices.

Finish: Dry and warm.

Notes

Limited edition. Finished an extra 21 days in toasted French oak ex-red wine barrels.

My Score

Notes: ______________________________

____

____

Appearance (1-5) ____ Nose (1-25) ____

Taste (1-25) ____ Finish (1-25) ____

Complexity (1-10) ____ Overall Impression (1-10) ____

SCORE ____

Bluegrass Wheated Bottled-in-Bond

Proof: 100
Age: 4 years
Type: Straight
Style: Wheated, Pot Distilled
Mash Bill: 75% corn, 21% wheat, 4% malted barley
Color: Dark Straw
Price: $$$

Mike

Nose: Caramel and oak with hints of apples and pears.
Taste: Caramel, corn, pepper spice, oak, and a hint of fruit.
Finish: Long with lots of oak leading to pepper spice.

Susan

Nose: Loads of grain, primarily corn, but wheat is here, too. A touch of fruit and brown sugar.
Taste: Vanilla and some honey with berries and lots of black pepper.
Finish: Long and peppery.

Notes

There's a 90 proof expression of the wheated, too.

My Score

Notes: ____________________________________
____
____

Appearance (1-5) ____ Nose (1-25) ____
Taste (1-25) ____ Finish (1-25) ____
Complexity (1-10) ____ Overall Impression (1-10) ____
SCORE ____

Boatwright Bourbon

Proof: 90
Age: NAS
Type: Straight
Style: Blend of two straight bourbons
Mash Bill: Not available
Color: Light Straw
Price: $$$

Mike

Nose: Vanilla, apples, baking spice, oak.
Taste: Vanilla, apples, cinnamon, nutmeg, oak.
Finish: Medium long with oak and cinnamon.

Susan

Nose: Vanilla, sweet oak, some cinnamon sugar.
Taste: Vanilla wafers, cinnamon, apple peel.
Finish: Medium long, warm, smooth.

Notes

Uses bourbon from Kentucky and Port Chilkoot Distillery, Haines, Alaska.

My Score

Notes: ______________________________________

____

____

Appearance (1-5) ____ Nose (1-25) ____
Taste (1-25) ____ Finish (1-25) ____
Complexity (1-10) ____ Overall Impression (1-10) ____

SCORE ____

Bodalen Bourbon

Proof: 94
Age: 18 months
Type: Not Straight
Style: Traditional
Mash Bill: 60% heirloom corn, 30% Hazlet rye, 10% malted barley
Color: Straw
Price: $

Mike

Nose: Caramel and toffee with oak and a hint of citrus.
Taste: Honey, baking spices, oak, and a little tobacco.
Finish: Long and spicy.

Susan

Nose: Multigrain cereal, vanilla, and sugar. Smells like a bakery!
Taste: Sweet oak and vanilla with a little citrus.
Finish: Medium long and oaky.

Notes

The distiller grows the grains on his family's farm. Pronounced BO-dah-lyn.

My Score

Notes: ____________________________________

____

____

Appearance (1-5) ____ Nose (1-25) ____
Taste (1-25) ____ Finish (1-25) ____
Complexity (1-10) ____ Overall Impression (1-10) ____
SCORE ____

Boondocks Port Barrel Finished

Proof:	90
Age:	8 years
Type:	Straight
Style:	Finished
Mash Bill:	Not available
Color:	Straw
Price:	$$$

Mike

Nose: Cherries and berries with vanilla, oak, and a hint of almonds.

Taste: Berries and vanilla, a little sweet baking spice, oak and a hint of that almond.

Finish: Long, dry with oak, spice and a hint of tart berries.

Susan

Nose: Very fruity. Bananas, peaches, even pineapple, with a bit of vanilla.

Taste: Banana nut bread with some caramel icing.

Finish: Extremely smooth. Nicely coats the tongue and lingers, especially the nuts.

Notes

Boondocks sources whiskeys for their releases. Look for other expression, too.

My Score

Notes: ______________________________

____

____

Appearance (1-5)	____	Nose (1-25)	____
Taste (1-25)	____	Finish (1-25)	____
Complexity (1-10)	____	Overall Impression (1-10)	____

SCORE ____

Bootlegger 21

Proof:	92
Age:	18 months
Type:	Not Straight
Style:	Small Batch
Mash Bill:	Not available
Color:	Straw
Price:	$$$

Mike

Nose: Bread dough, yeast, hint of corn and vanilla.
Taste: Bread dough, vanilla, hint of pepper.
Finish: Short and peppery.

Susan

Nose: Corn and bread dough. Nut shells.
Taste: Like chewing on pistachios shells with some vanilla.
Finish: Short and warm.

Notes

Prohibition Distillery, Roscoe, New York. Batch #2, Bottle 453. Sold in 375ml bottles.

My Score

Notes: ____________________________________

____

____

Appearance (1-5) ____ Nose (1-25) ____
Taste (1-25) ____ Finish (1-25) ____
Complexity (1-10) ____ Overall Impression (1-10) ____

SCORE ____

Boulder Spirits Bourbon 2-Year-Old

Proof: 84
Age: 2 years
Type: Straight
Style: Pot Distilled
Mash Bill: 51% corn, 5% rye, 44% malted barley
Color: Light Straw
Price: $$

Mike

Nose: Vanilla, banana, citrus, oak.
Taste: Vanilla, banana, white pepper, oak.
Finish: Medium long with oak and pepper.

Susan

Nose: Caramel, pear, baking spice, sweet oak.
Taste: Brown sugar, cinnamon sugar cookies, pears. Well balanced.
Finish: Dries to light oak.

Notes

Vapor Distilling, Boulder, Colorado. One of many expressions produced.

My Score

Notes: ____

____

____

Appearance (1-5) ____ Nose (1-25) ____
Taste (1-25) ____ Finish (1-25) ____
Complexity (1-10) ____ Overall Impression (1-10) ____

SCORE ____

Boulder Spirits Bourbon 6-Year-Old

Proof: 115.2
Age: 6 years
Type: Straight
Style: Pot Distilled, Small Batch
Mash Bill: 51% corn, 5% rye, 44% malted barley
Color: Dark Straw
Price: $$$

Mike

Nose: Vanilla, pears, baking spice, oak.
Taste: Vanilla, pears, black pepper, oak.
Finish: Long and dry with oak and pepper.

Susan

Nose: Pears, honey, light oak, dark brown sugar, touch of citrus.
Taste: Cinnamon candy with honey, cherries, caramel.
Finish: All flavor notes fade in tandem, but pear eventually wins.

Notes

Vapor Distilling, Boulder, Colorado. One of many expressions produced. Limited release. Only 1,800 bottles.

My Score

Notes: ______________________________

____

____

Appearance (1-5) ____ Nose (1-25) ____
Taste (1-25) ____ Finish (1-25) ____
Complexity (1-10) ____ Overall Impression (1-10) ____

SCORE ____

Bourbon 30 Barrel Crafted

Proof:	90
Age:	6 months
Type:	Not Straight
Style:	Small Batch
Mash Bill:	Not available
Color:	Straw
Price:	$$$

Mike

Nose: Young, but citrus notes and vanilla. Corn and a hint of oak.

Taste: Thin in the mouth. Caramel and corn with a hint of oak.

Finish: Thin and short with just a hint of oak.

Susan

Nose: Maple syrup, caramel, dash of cinnamon.

Taste: Corn, some honey. White pepper.

Finish: Short. Dry oak ending.

Notes

Distilled and bottled by Three Boys Farm Distillery, Graefenburg, Kentucky.

My Score

Notes: ____

____

____

Appearance (1-5) ____ Nose (1-25) ____

Taste (1-25) ____ Finish (1-25) ____

Complexity (1-10) ____ Overall Impression (1-10) ____

SCORE ____

Bourbon 30 Single Barrel

Proof:	119
Age:	4 months
Type:	Not Straight
Style:	Special Finish, Barrel Strength, Single Barrel
Mash Bill:	Not available
Color:	Straw
Price:	$$$

Mike

Nose: Corn, pears and berries, vanilla and oak.
Taste: Fruity with berries and vanilla, oak, an spice.
Finish: Starts fruity, then dries to oak and spice.

Susan

Nose: Bacon and wood smoke.
Taste: Pears, vanilla, some sweet spices.
Finish: Short, peppery.

Notes

From Three Boys Farm Distillery, Graefenburg, Kentucky. Finished with toasted barrel staves. Barrel #14, bottle 195.

My Score

Notes: ____

____

____

Appearance (1-5) ____ Nose (1-25) ____
Taste (1-25) ____ Finish (1-25) ____
Complexity (1-10) ____ Overall Impression (1-10) ____

SCORE ____

Bourbon 30 Small Batch

Proof: 118.5
Age: 4 months
Type: Not Straight
Style: Small Batch, Barrel Strength
Mash Bill: Not available
Color: Straw
Price: $$$

Mike

Nose: Pears and citrus with corn and vanilla.
Taste: Corn, pears and vanilla. Hint of oak.
Finish: Dry with oak and pepper.

Susan

Nose: Light. Some corn and citrus.
Taste: Corn with some pear. Water releases some sweet spice.
Finish: Very peppery.

Notes

Bottle #524, batch #11. Made by Three Boys Farm Distillery, Frankfort, Kentucky.

My Score

Notes: ____________________________________

____

____

Appearance (1-5) ___ Nose (1-25) ___
Taste (1-25) ___ Finish (1-25) ___
Complexity (1-10) ___ Overall Impression (1-10) ___

SCORE ___

Bower Hill Barrel Reserve

Proof: 86
Age: NAS
Type: Straight
Style: Small Batch
Mash Bill: Not available
Color: Straw
Price: $$

Mike

Nose: Bread dough and yeast. Some corn and vanilla.
Taste: Corn bread and vanilla. Hint of pepper.
Finish: Short and sweet with hint of oak.

Susan

Nose: Caramel corn, honey, sweet spices.
Taste: More caramel corn, some ripe peach.
Finish: Pepper and oak.

Notes

Distilled and aged in Louisville, Kentucky. Bottles by Bower Hill Distillery, Silverton, Ohio. Bower Hill was the deadliest battle of the Whiskey Rebellion of 1794. There is also a very limited barrel strength release.

My Score

Notes: ______________________________

____

____

Appearance (1-5) ____ Nose (1-25) ____
Taste (1-25) ____ Finish (1-25) ____
Complexity (1-10) ____ Overall Impression (1-10) ____

SCORE ____

Breaker

Proof: 90
Age: 5 years
Type: Straight
Style: Small Batch
Mash Bill: Not available
Color: Straw
Price: $$$

Mike

Nose: Corn, vanilla, sweet spices. Hint of peaches.
Taste: Corn and vanilla with some peaches and oak.
Finish: Starts sweet and dries with some oak.

Susan

Nose: Light caramel, apples and pears.
Taste: Caramel and a bit of chocolate with some apples.
Finish: Dry and oaky.

Notes

Distilled by MGP. Bottled by Ascendant Spirits, Santa Barbara Co., California. Batch a mingling of 8 barrels. This was bottle 774 from batch 13. More recent releases *may* be distilled in California.

My Score

Notes: ______________________________

____

____

Appearance (1-5) ____ Nose (1-25) ____
Taste (1-25) ____ Finish (1-25) ____
Complexity (1-10) ____ Overall Impression (1-10) ____
SCORE ____

Breaking & Entering

Proof: 86
Age: NAS
Type: Not Straight
Style: Small Batch
Mash Bill: Not available
Color: Light Straw
Price: $$$

Mike

Nose: Vanilla and corn with floral notes and fruit – apricots and berries.
Taste: Sweet caramel, light apricot, and some leather. Bit of spice.
Finish: Very long and dry with oak and pepper.

Susan

Nose: Vanilla, pitted fruit, a little honey and light spice.
Taste: Vanilla, oak, a little nutmeg.
Finish: Medium long and oaky. Dash of pepper.

Notes

St. George Spirits. Alameda, California. Craft distillery, but this is a batching of some 80 barrels of different bourbons sourced from Kentucky. Distilleries unknown.

My Score

Notes: ____________________

____

____

Appearance (1-5) ____ Nose (1-25) ____
Taste (1-25) ____ Finish (1-25) ____
Complexity (1-10) ____ Overall Impression (1-10) ____

SCORE ____

Breckenridge

Proof: 86
Age: 2-3 years
Type: Straight
Style: Small Batch, Craft
Mash Bill: 56% corn, 38% rye, 6% malted barley
Color: Dark Straw
Price: $$$

Mike

Nose: Corn and oak with a hint of spice. A little fruit in the background
Taste: Corn and vanilla with a hint of spice. A bit thin and watery. Not complex.
Finish: Short and sweet.

Susan

Nose: Orange peel, vanilla, cinnamon.
Taste: Pepper on the palate overwhelms any other flavors.
Finish: Short, but not hot.

Notes

From Breckenridge Distillery, Breckenridge, Colorado. Earlier expressions of the bourbon were sourced from Kentucky. Own expression released in 2014. Bourbon is proofed using snowmelt.

My Score

Notes: ____________________________________

____

____

Appearance (1-5) ____ Nose (1-25) ____
Taste (1-25) ____ Finish (1-25) ____
Complexity (1-10) ____ Overall Impression (1-10) ____
SCORE ____

Breckenridge High Proof Blend

Proof:	105
Age:	Minimum 2 years
Type:	Straight
Style:	Blend
Mash Bill:	Not revealed
Color:	Straw
Price:	$$$

Mike

Nose: Very young. Corn with a hint of vanilla and oak.

Taste: Very light for 105 proof. Vanilla with a hint of citrus and spice.

Finish: Medium long with just a hint of oak and some baking spice.

Susan

Nose: Sweet mint, vanilla, some citrus and banana.

Taste: Banana nut bread with an odd rubbery note. Sweeten with a little water and the odd vanishes.

Finish: Peppery bordering on hot.

Notes

If you visit, know there is an excellent restaurant at the distillery. Breckenridge Distillery Breckenridge, Colorado, https://breckenridgedistillery.com.

My Score

Notes: ____

____

____

Appearance (1-5) ____ Nose (1-25) ____

Taste (1-25) ____ Finish (1-25) ____

Complexity (1-10) ____ Overall Impression (1-10) ____

SCORE ____

Broken Bell

Proof: 90
Age: NAS
Type: Not Straight
Style: Small Batch
Mash Bill: Not available
Color: Light Straw
Price: $$

Mike

Nose: Very light. A hint of vanilla and corn.
Taste: Very light on the tongue with a hint of candy corn and spice.
Finish: Gets peppery with time and lasts longer than taste would indicate.

Susan

Nose: Vanilla, but very light. Perhaps a bit of cinnamon or nutmeg.
Taste: Mirrors the nose. Some vanilla and fairly full mouthfeel. Not complex.
Finish: Medium long and dry.

Notes

Broken Bell Whiskey Co., Princeton, Minnesota. Sourced from an unidentified distillery. Does not say "Kentucky" on the label.

My Score

Notes: ____
____
____

Appearance (1-5) ____ Nose (1-25) ____
Taste (1-25) ____ Finish (1-25) ____
Complexity (1-10) ____ Overall Impression (1-10) ____
SCORE ____

Brush Creek Bourbon

Proof: 94
Age: 4-5 years
Type: Straight
Style: Blend
Mash Bill: Not revealed
Color: Light Straw
Price: $$

Mike

Nose: Light nose. Vanilla, hint of apples and spice. Oak.
Taste: Vanilla, apples, a little pepper spice, and oak.
Finish: Lots of pepper spice and a little oak.

Susan

Nose: Buttered popcorn, nuts, and a whiff of orange.
Taste: Some caramel and vanilla with light orange peel and peppery oak.
Finish: Spicy without being hot.

Notes

Balanced but not complex. Currently sourcing and blending bourbons from Kentucky, Indiana, and Tennessee while their own whiskeys mature. Brush Creek Distillery, Saratoga, Wyoming, www.brushcreekdistillery.com.

My Score

Notes: ____________________________________

____

____

Appearance (1-5) ____ Nose (1-25) ____
Taste (1-25) ____ Finish (1-25) ____
Complexity (1-10) ____ Overall Impression (1-10) ____

SCORE ____

Buck

Proof: 90
Age: 8 years
Type: Straight
Style: Small Batch
Mash Bill: Not available
Color: Dark Straw
Price: $$

Mike

Nose: Very light. Some vanilla and saddle leather.
Taste: Very light. Some apple and caramel leading to a spicy finish.
Finish: Fairly short and peppery.

Susan

Nose: Cinnamon, oak, nuts.
Taste: Ripe pears, vanilla and caramel. A dash of black pepper.
Finish: Medium long and tannic.

Notes

Frank-Lin Distillers, Fairfield, California. This facility sources bourbon from undisclosed distilleries in Kentucky and bottles it under several labels. This is curiously simple bourbon for 8 years old.

My Score

Notes: ______________________________

____

____

Appearance (1-5) ____ Nose (1-25) ____
Taste (1-25) ____ Finish (1-25) ____
Complexity (1-10) ____ Overall Impression (1-10) ____

SCORE ____

Burks Spring Bourbon

Proof: 100
Age: 4 years
Type: Straight
Style: Bottled-in-Bond
Mash Bill: 60% corn, 25% rye, 15% malted barley
Color: Light Amber
Price: $$$

Mike

Nose: Vanilla, with a hint of maple. Ripe apples, oak.
Taste: Vanilla, maple, apples, nutmeg, allspice, oak.
Finish: Medium long with oak and nutmeg.

Susan

Nose: Light apple peel, herbal spices, vanilla.
Taste: Brown sugar, nut, baking spice, and ripe apple.
Finish: Long and balanced between sweet and savory.

Notes

Distilled by Kentucky Artisan Distillers, Crestwood, Kentucky. Bottled by Burks Spring Distilling Company, Loretto, Kentucky.

My Score

Notes: ____________________________________

____

____

Appearance (1-5) ____ Nose (1-25) ____
Taste (1-25) ____ Finish (1-25) ____
Complexity (1-10) ____ Overall Impression (1-10) ____

SCORE ____

Burnside Double Barrel

Proof: 96
Age: NAS
Type: Straight
Style: Special Finish
Mash Bill: 75% corn, 10% rye, 15% malted barley
Color: Straw
Price: $$$

Mike

Nose: Sweet vanilla and berries with some ginger.
Taste: Sweet vanilla and ginger with some citrus and tobacco.
Finish: Starts sweet and dries to oak tannins.

Susan

Nose: Toasted marshmallow and light fruit.
Taste: Vanilla, ginger, and orange rind.
Finish: Sweet vanilla ending in spicy oak.

Notes

From Eastside Distilling, Portland, Oregon. The company sources its mashed grain (No word from where.) and ferments, distills, and bottles. After initial aging, this is aged an extra 60 days in heavily charred, Oregon oak barrels. This was Batch No. 6, Bottle No. 325.

My Score

Notes: ____________________

____

____

Appearance (1-5) ____ Nose (1-25) ____
Taste (1-25) ____ Finish (1-25) ____
Complexity (1-10) ____ Overall Impression (1-10) ____

SCORE ____

Buzzard's Roost Barrel Strength

Proof:	114.4
Age:	4-6 years
Type:	Straight
Style:	Small Batch, Secondary Barrel, Cask Strength
Mash Bill:	Blend of the 36% rye and 21% rye mash bills
Color:	Dark Straw
Price:	$$$

Mike

Nose: Vanilla and fruit – apples and pears. A little baking spice and oak.

Taste: Vanilla and fruit – lots of blackberries and raspberries a with a hint of pears. Baking spices and oak.

Finish: Long with oak and spice and some lingering fruit.

Susan

Nose: Baking cornbread, brown sugar, ripe peaches, and some faint nuts and oak.

Taste: Vanilla pudding sprinkled with dark brown sugar, candied almonds and some berries balanced with baking spices.

Finish: Long and spicy.

Notes

Very complex. Sourced from MGP in Indiana and secondarily aged in proprietary barrels.

My Score

Notes: ______________________________

____

____

Appearance (1-5) ____ Nose (1-25) ____

Taste (1-25) ____ Finish (1-25) ____

Complexity (1-10) ____ Overall Impression (1-10) ____

SCORE ____

Buzzard's Roost Bottled-in-Bond

Proof:	100
Age:	5 years
Type:	Straight
Style:	Bottled-in-Bond, Wheated
Mash Bill:	51% corn, 45% wheat, 4% malted barley
Color:	Dark Straw
Price:	$$$

Mike

Nose: Vanilla, dates, marshmallows, oak.
Taste: Vanilla, cherries, dates, black pepper, and oak.
Finish: Long with oak and pepper.

Susan

Nose: Apples, pears, vanilla wafers, brown sugar, some honey, and citrus.
Taste: Rich vanilla balanced with caramel, ripe cherries, and baking spices.
Finish: Lots of fruit and spice linger until fading into light pepper and oak.

Notes

Balanced and complex. Released in 2024 and a welcome addition to the bonded bourbons on the market.

My Score

Notes: ______________________________________

____

____

Appearance (1-5) ____ Nose (1-25) ____
Taste (1-25) ____ Finish (1-25) ____
Complexity (1-10) ____ Overall Impression (1-10) ____

SCORE ____

Buzzard's Roost Char #1

Proof: 105
Age: NAS
Type: Straight
Style: Small Batch, Secondary Barrel
Mash Bill: 59% corn, 36% rye, 5% malted barley
Color: Straw
Price: $$$

Mike

Nose: Vanilla and butterscotch with a hint of apples and oak.
Taste: Vanilla, ripe apples, allspice, and oak.
Finish: Medium long with oak and spice.

Susan

Nose: Caramel, honey, roasted corn, light cinnamon, vanilla pudding, some apple/pear.
Taste: Cola, honey, caramel, apples, cinnamon.
Finish: Long, drying to oak.

Notes

Buzzard's Roost Spirits, Louisville, Kentucky. Originally sourced from MPG and secondarily aged in custom, proprietary barrels. Now contract distilled and still aged in custom-heated barrels.

My Score

Notes: ______________________________

____

____

Appearance (1-5) ____ Nose (1-25) ____
Taste (1-25) ____ Finish (1-25) ____
Complexity (1-10) ____ Overall Impression (1-10) ____

SCORE ____

Buzzard's Roost Toasted American Oak

Proof: 105
Age: 4 and 5 years
Type: Straight
Style: Small Batch, Secondary Barrel
Mash Bill: 280 of 74% corn, 21% rye, 5% malted barley and 59% corn, 36% rye, 5% malted barley
Color: Light Straw
Price: $$$

Mike

Nose: Caramel apples, baking spice, and oak.
Taste: Caramel, apples, white pepper, oak.
Finish: Medium long with oak and pepper.

Susan

Nose: Light brown sugar, crème brulee, apples, light cinnamon.
Taste: Vanilla pudding, red currents, nutmeg, dash of cinnamon.
Finish: Dries to sweet oak.

Notes

Buzzard's Roost Spirits, Louisville, Kentucky. Look for additional, limited release expressions, too, such as Toasted French Oak.

My Score

Notes: ______________________________

____

____

Appearance (1-5) ____ Nose (1-25) ____
Taste (1-25) ____ Finish (1-25) ____
Complexity (1-10) ____ Overall Impression (1-10) ____
SCORE ____

Calumet Farm

Proof: 86
Age: NAS
Type: Does not say "Straight" on label
Style: Traditional
Mash Bill: Not available
Color: Light Straw
Price: $$$

Mike

Nose: Very floral – honeysuckle and vanilla sweet corn and some oak.
Taste: Sweet corn and vanilla up front with pepper spice coming afterwards.
Finish: Long and dry with oak.

Susan

Nose: Very light. Some caramel and dried fruit.
Taste: Light maple syrup. Some caramel.
Finish: Long and dry. Best aspect of this bourbon.

Notes

Kentucky-sourced bourbon of indeterminate origin. Bottled by Three Springs Bottling Co. of Bowling Green and owned by Western Spirits of the same city. Pretty packaging and the famed Calumet Farm name (producer of eight Kentucky Derby winners) are the rationales for the high price tag. Contents not unpleasant, but should be much better for the money.

My Score

Notes: ______________________________

____

____

Appearance (1-5) ____ Nose (1-25) ____
Taste (1-25) ____ Finish (1-25) ____
Complexity (1-10) ____ Overall Impression (1-10) ____
SCORE ____

Calumet Farm Single Rack Black

Proof: 92
Age: 10 years
Type: Straight
Style: Small Batch
Mash Bill: 74% corn, 18% rye, 8% malted barley
Color: Straw
Price: $$$$

Mike

Nose: Vanilla and sweet cherry bubble gum with a hint of oak.
Taste: Corn, vanilla, a bit of fruit – berries and cherries – with light baking spice and oak. Thin mouthfeel.
Finish: Medium long with some oak.

Susan

Nose: Caramel apples and some milk chocolate.
Taste: Chocolate-coated caramel, some cherries and cinnamon. Thin mouthfeel.
Finish: Short with just a little caramel.

Notes

Nineteen barrel batches. One of several Rack Black limited releases. Look for others of various ages.

My Score

Notes: ____________________________________

____

____

Appearance (1-5) ____ Nose (1-25) ____
Taste (1-25) ____ Finish (1-25) ____
Complexity (1-10) ____ Overall Impression (1-10) ____
SCORE ____

Cassius Clay

Proof: 100
Age: NAS
Type: Straight
Style: Traditional
Mash Bill: Not available
Color: Light Straw
Price: $$$

Mike

Nose: Corn and orange blossom. Honey.
Taste: Corn and honey with some peaches and oak.
Finish: Long. Starts with sweet peaches and vanilla. Dries to oak.

Susan

Nose: Caramel corn, raisins.
Taste: Sweet vanilla and some berries.
Finish: Warm and sweet with oak at the end.

Notes

Not complex, but smooth and enjoyable. Distilled in Tennessee and bottled in Bardstown, Kentucky by Cary King Bottling Co.

My Score

Notes: ____
____
____

Appearance (1-5) ____ Nose (1-25) ____
Taste (1-25) ____ Finish (1-25) ____
Complexity (1-10) ____ Overall Impression (1-10) ____

SCORE ____

Castle & Key Small Batch

Proof:	100
Age:	4 years
Type:	Straight
Style:	Traditional
Mash Bill:	73% white corn, 10% rye, 17% malted barley
Color:	Straw
Price:	$$$

Mike

Nose: Very fruit forward with vanilla, marshmallows, wild berries, plums, oakwood.
Taste: Vanilla, blackberries, black pepper, and oak.
Finish: Medium long with oak and pepper.

Susan

Nose: Fruit cocktail syrup, brown sugar, nutmeg, vanilla.
Taste: Caramel, citrus, berries, and black pepper.
Finish: Medium long and peppery.

Notes

Balanced and complex. Released in 2024 and a welcome addition to the bonded bourbons on the market.

My Score

Notes: ______________________________

____

____

Appearance (1-5) ___ Nose (1-25) ___
Taste (1-25) ___ Finish (1-25) ___
Complexity (1-10) ___ Overall Impression (1-10) ___

SCORE ___

Catamount

Proof: 88
Age: NAS
Type: Straight
Style: Wheated
Mash Bill: 51% corn, 45% wheat, 4% malted barley
Color: Straw
Price: $$

Mike

Nose: Vanilla and biscuit dough with a little apple and oak.
Taste: Vanilla and ripe apples with some pepper and oak.
Finish: Long and peppery.

Susan

Nose: Bananas and apples with some vanilla.
Taste: Vanilla, light milk chocolate with some apple.
Finish: Long with some spice.

Notes

A little water make it even fruitier. Grand Teton Distillery, Driggs, Idaho. https://www.tetondistillery.com.

My Score

Notes: ______________________________

____

____

Appearance (1-5) ____ Nose (1-25) ____
Taste (1-25) ____ Finish (1-25) ____
Complexity (1-10) ____ Overall Impression (1-10) ____

SCORE ____

Cedar Ridge

Proof: 80
Age: NAS
Type: Not Straight
Style: Small Batch
Mash Bill: ~ 75% corn
Color: Very Light Straw
Price: $$$

Mike

Nose: Corn and vanilla. very light.
Taste: Thin and watery. Corn and vanilla
Finish: Short and sweet with vanilla.

Susan

Nose: Corn and cellulose.
Taste: Vanilla and sweet corn. Not complex, but pleasant.
Finish: Very warm for 80 proof. Lingers.

Notes

Cedar Ridge Distillery, Swisher, Iowa. Made in copper pot stills.

My Score

Notes: ______________________________

____

____

Appearance (1-5) ____ Nose (1-25) ____
Taste (1-25) ____ Finish (1-25) ____
Complexity (1-10) ____ Overall Impression (1-10) ____

SCORE ____

CH Bourbon

Proof:	105
Age:	NAS
Type:	Not Straight
Style:	Traditional
Mash Bill:	Not available
Color:	Straw
Price:	$$$

Mike

Nose: Corn, caramel and a hint of spice.
Taste: Corn and vanilla with some ripe pear and baking spice.
Finish: Starts sweet and dries nicely.

Susan

Nose: Corn and sweet alcohol.
Taste: Light vanilla, light cornbread, some apples.
Finish: Warm with pepper notes.

Notes

Sourced by CH Distillery, Chicago, Illinois, which distills gin, vodka and rum.

My Score

Notes: ______________________________

____

____

Appearance (1-5) ___ Nose (1-25) ___
Taste (1-25) ___ Finish (1-25) ___
Complexity (1-10) ___ Overall Impression (1-10) ___

SCORE ___

Charles Goodnight

Proof:	100
Age:	NAS
Type:	Straight
Style:	Small Batch
Mash Bill:	Not available
Color:	Straw
Price:	$$$

Mike

Nose: Vanilla and corn. Very light.
Taste: Rotten corn and vanilla with hint of oak.
Finish: Dry with spoiled corn.

Susan

Nose: Corn with some vanilla sweetness.
Taste: Overripe corn. Some berries and vanilla.
Finish: Short and dry. Funky aftertaste.

Notes

Small batch #0012. Named after a famous Texas Ranger and cattleman. Sourced from Kentucky and bottled by Goodnight Distilling, Parlier, California.

My Score

Notes: ______________________________

____

____

Appearance (1-5) ____ Nose (1-25) ____
Taste (1-25) ____ Finish (1-25) ____
Complexity (1-10) ____ Overall Impression (1-10) ____
SCORE ____

Charred Oak

Proof: 100
Age: NAS
Type: Not Straight
Style: Traditional
Mash Bill: Not available
Color: Light Straw
Price: $$

Mike

Nose: Vanilla and corn. Very light.
Taste: Corn and pepper spice. Hint of corn silk.
Finish: Spicy with hint of oak and corn silk.

Susan

Nose: Alcohol with some corn and vanilla.
Taste: Nail polish.
Finish: Hot, dry.

Notes

No indication on the label of the origin of this bourbon. Website states "blend of Kentucky and Wisconsin bourbons." Blended and bottled by Off-the-Clock Spirits, Madison, Wisconsin.

My Score

Notes: ______________________________

____

____

Appearance (1-5) ____ Nose (1-25) ____
Taste (1-25) ____ Finish (1-25) ____
Complexity (1-10) ____ Overall Impression (1-10) ____

SCORE ____

Chattanooga Whiskey

Proof: 111
Age: 2 years
Type: Straight
Style: Small Batch, Non-Chill Filtered, Four Grain, Barrel Strength
Mash Bill: 75% yellow corn, 25% malted rye, caramel malt, honey malt
Color: Light Straw
Price: $$

Mike

Nose: Very sweet. Vanilla, honey, wild berries, and oak.
Taste: Corn and malt, honey, blackberry, oak.
Finish: Short with oak and pepper.

Susan

Nose: Slightly burnt caramel, apple peel, some vanilla and honey.
Taste: Dark fruit, caramel, leather and oak with a dash of pepper.
Finish: Medium long and peppery.

Notes

Chattanooga Whiskey Company, Tennessee. Small batch is 6-10 barrels. Look for other, lower proof, expressions, too.

My Score

Notes: ____________________________________

____

____

Appearance (1-5) ____ Nose (1-25) ____
Taste (1-25) ____ Finish (1-25) ____
Complexity (1-10) ____ Overall Impression (1-10) ____

SCORE ____

Chestnut Farms

Proof: 90
Age: NAS
Type: Straight
Style: Traditional
Mash Bill: Not available
Color: Light Amber
Price: $$$

Mike

Nose: Caramel and vanilla, corn and leather.
Taste: Corn, vanilla and pepper with hint of tobacco.
Finish: Short and dry with oak and pepper.

Susan

Nose: Spicy with vanilla, caramel and baked apples.
Taste: Some orange peel, baking spices, caramel.
Finish: Warm and oaky with pepper at the end.

Notes

From Clear Springs Distilling Co., Louisville, Kentucky.

My Score

Notes: ____________________________________
____
____

Appearance (1-5) ___ Nose (1-25) ___
Taste (1-25) ___ Finish (1-25) ___
Complexity (1-10) ___ Overall Impression (1-10) ___

SCORE ___

Chicken Cock 160th Anniversary

Proof: 90
Age: 8 years
Type: Straight
Style: Single Barrel
Mash Bill: 70% corn, 21% wheat, 9% malted barley
Color: Straw
Price: $$$$

Mike

Nose: Toasted marshmallows, a little fruit and oak.
Taste: Vanilla, corn, apples and pears, with some pepper spice and oak.
Finish: Medium long with oak and spice.

Susan

Nose: Burnt toffee, figs, and dates, with a dash of cinnamon.
Taste: Caramel and dark fruit with plenty of pepper and oak.
Finish: Long, spicy, and dry.

Notes

Sourced. Limited edition but can be found in large bourbon bars. Touch of water tames the heat.

My Score

Notes: ____________________________________

____

____

Appearance (1-5) ____ Nose (1-25) ____
Taste (1-25) ____ Finish (1-25) ____
Complexity (1-10) ____ Overall Impression (1-10) ____

SCORE ____

Chicken Cock Heritage Reserve

Proof: 90
Age: 6 months
Type: Not Straight
Style: Small Batch
Mash Bill: Not available
Color: Straw
Price: $$$

Mike

Nose: Yeast, corn and hint of peaches and vanilla.
Taste: Brown vodka. Not much flavor – corn and a little oak.
Finish: Dry with alcohol and faintest hint of oak.

Susan

Nose: Ripe apples, berries with vanilla.
Taste: Faint, but apples, vanilla and some banana.
Finish: Short with pepper.

Notes

This brand originated in the 1850s, but went extinct. Grain and Barrel of North Charleston, South Carolina, acquired the name and is sourcing and bottling

My Score

Notes: ____

____

____

Appearance (1-5) ____ Nose (1-25) ____
Taste (1-25) ____ Finish (1-25) ____
Complexity (1-10) ____ Overall Impression (1-10) ____

SCORE ____

Cleveland Bourbon

Proof: 87
Age: Less than 2 years
Type: Not Straight
Style: Age Accelerated
Mash Bill: Not available
Color: Very Light Straw
Price: $$$

Mike

Nose: Corn and oak.
Taste: Corn and oak and not much else.
Finish: Oak and lots of it.

Susan

Nose: Corn and wood.
Taste: Sweet corn and alcohol.
Finish: Sweet oak.

Notes

Distilled in Indiana, but pressure aged and bottled by Cleveland Whiskey, LLC., Ohio. So something of a hybrid rather than true craft distiller. Slogan: "Drink what you like. Age is irrelevant."

My Score

Notes: ____________________________________
____
____

Appearance (1-5) ____ Nose (1-25) ____
Taste (1-25) ____ Finish (1-25) ____
Complexity (1-10) ____ Overall Impression (1-10) ____

SCORE ____

Cody Road

Proof: 90
Age: Less than 4 years
Type: Not Straight
Style: Small Batch
Mash Bill: Not available
Color: Very Pale Straw
Price: $$

Mike

Nose: Very light. Sweet vanilla and peaches.
Taste: Very light, but pleasant. Peaches and vanilla with a hint of spice.
Finish: Short and simple. Just a hint of oak.

Susan

Nose: Corn and a little sweet spice.
Taste: Rich mouthfeel, but not complex. Some sweet corn and a little sweet oak.
Finish: Smooth and sweet. Not hot.

Notes

Mississippi River Distilling Co., LeClaire, Iowa. This was Batch 7, Bottle 273. Would be interesting to see what some more aging would do for this.

My Score

Notes: ____________________________________

____

____

Appearance (1-5) ____ Nose (1-25) ____
Taste (1-25) ____ Finish (1-25) ____
Complexity (1-10) ____ Overall Impression (1-10) ____

SCORE ____

Colter's Run Single Barrel

Proof: 88
Age: NAS
Type: Straight
Style: Single Barrel
Mash Bill: 74% corn, 21% rye, 5% malted barley
Color: Light Straw
Price: $$

Mike

Nose: Very light. Vanilla and leather with a hint of oak.
Taste: Vanilla and a little orange peel, cardamom, and white pepper. A little oak.
Finish: Short with some oak and spice.

Susan

Nose: Very light with a whiff of hard candy and sweet new leather.
Taste: Caramel candy, some light spice and light apricot.
Finish: Bit of a tingle drying to oak.

Notes

Nice balance, but not complex. Barrel 58, bottle 159. Sourced from MGP in Indiana and aged in Idaho.

My Score

Notes: ______________________________

____

____

Appearance (1-5) ____ Nose (1-25) ____
Taste (1-25) ____ Finish (1-25) ____
Complexity (1-10) ____ Overall Impression (1-10) ____
SCORE ____

Colter's Run Small Batch

Proof:	88
Age:	Minimum 3 years
Type:	Straight
Style:	Small Batch
Mash Bill:	74% corn, 21% rye, 5% malted barley
Color:	Straw
Price:	$$

Mike

Nose: Vanilla and pipe tobacco with a hint of oak.

Taste: Vanilla and apricots with some cinnamon spice, tobacco, and oak.

Finish: Medium long with oak and tobacco.

Susan

Nose: Sweet mint, buttered corn, some berries.

Taste: More buttered corn, a little cotton candy, sweet cherries and a whiff of sweet tobacco.

Finish: Warm and dry with a dash of pepper.

Notes

By all means, try it with a pipe or cigar. Sourced from MGP in Indiana and aged in Idaho. Grand Teton Distillery, Driggs, Idaho. A little water makes it even fruitier. https://www.tetondistillery.com.

My Score

Notes: ______________________________

____

____

Appearance (1-5) ____ Nose (1-25) ____

Taste (1-25) ____ Finish (1-25) ____

Complexity (1-10) ____ Overall Impression (1-10) ____

SCORE ____

Conviction Cask Strength

Proof:	125.8
Age:	NAS
Type:	Straight
Style:	Barrel Strength
Mash Bill:	Not available
Color:	Light Amber
Price:	$$

Mike

Nose: Vanilla, chocolate, dates, oak.

Taste: Vanilla, milk chocolate, dates, wild berries, cinnamon, oak.

Finish: Long with oak, cinnamon, and lingering chocolate.

Susan

Nose: Medley of dark fruit – raisins, dates, figs – toffee, and savory spice.

Taste: Toffee bar studded with dates and sprinkled with grated cinnamon.

Finish: Long and smooth with a nice chocolate note at the end.

Notes

Southern Grace Distilleries, Mt. Pleasant, North Carolina. Housed in a former prison. Limited release, but look got it in well stocked whiskey bars. Aged in French oak casks.

My Score

Notes: ________________________________

____

____

Appearance (1-5) ____ Nose (1-25) ____

Taste (1-25) ____ Finish (1-25) ____

Complexity (1-10) ____ Overall Impression (1-10) ____

SCORE ____

Conviction Single Barrel Bourbon

Proof: 100
Age: 2 years
Type: Straight
Style: Single Barrel
Mash Bill: 75% corn, 21% rye, 4% malted barley
Color: Straw
Price: $$

Mike

Nose: Sweet vanilla, apricots, and baking spices.
Taste: Vanilla, apricots with a hint of berries, sweet baking spices, and oak.
Finish: Medium long with oak and spice.

Susan

Nose: Fruit salad with a lot of vanilla and caramel.
Taste: Sweet and fruity with vanilla. Grain reveals itself on the back of the tongue.
Finish: Dry and smooth with oak.

Notes

Barrel 76, bottle 145. The distillery is housed in a former prison. Other expressions include a 4-year, 105 proof bourbon. Southern Grace Distilleries, Mount Pleasant, North Carolina, https://www.southerngracedistilleries.com.

My Score

Notes: ______________________________

____

____

Appearance (1-5) ____ Nose (1-25) ____
Taste (1-25) ____ Finish (1-25) ____
Complexity (1-10) ____ Overall Impression (1-10) ____

SCORE ____

Conviction Small Batch Bourbon

Proof:	97.42
Age:	~2 years
Type:	Straight
Style:	Small Batch
Mash Bill:	75% corn, 21% rye, 4% malted barley
Color:	Very, Very Pale Straw
Price:	$$

Mike

Nose: Very light. Vanilla and corn with a hint of apricot jam.
Taste: Very light with the same vanilla, corn, and apricot jam.
Finish: Short and only a hint of oak.

Susan

Nose: Floral with lots of sugar.
Taste: Honey and nutmeg with vanilla and some oak.
Finish: Dries to pepper spice.

Notes

No need for water or ice, even at this proof. Southern Grace Distilleries, Mount Pleasant, North Carolina, https://www.southerngracedistilleries.com.

My Score

Notes: ______________________________

____

____

Appearance (1-5) ____ Nose (1-25) ____
Taste (1-25) ____ Finish (1-25) ____
Complexity (1-10) ____ Overall Impression (1-10) ____

SCORE ____

Cook's Mill

Proof: 88
Age: At least 2 years
Type: Straight
Style: Traditional
Mash Bill: Not available
Color: Light Straw
Price: $$$

Mike

Nose: Very light nose of vanilla and oak.
Taste: Vanilla, blackberries, pears, white pepper, oak.
Finish: Medium long with oak and pepper.

Susan

Nose: Light brown sugar, canned fruit cocktail, very light spice.
Taste: Caramel with some dried apple, nuts, tannins.
Finish: Warm with white pepper.

Notes

Queen Spirits, LLC. Statesville, North Carolina. Balanced, but not complex.

My Score

Notes: ____

____

____

Appearance (1-5) ____ Nose (1-25) ____
Taste (1-25) ____ Finish (1-25) ____
Complexity (1-10) ____ Overall Impression (1-10) ____
SCORE ____

Copper Fiddle

Proof: 92
Age: NAS
Type: Not Straight
Style: Traditional
Mash Bill: Not available
Color: Straw
Price: $$$

Mike

Nose: Fruity. Berries and honey with some spice.
Taste: Very fruity with butterscotch. Peaches, spice, a hint of oak.
Finish: Sweet caramel and oak.

Susan

Nose: Caramel and pecans. Dates.
Taste: Wood smoke, caramel, vanilla, apples.
Finish: Sweet, drying to oak and smoke.

Notes

Copper Fiddle Distillery, Lake Zurich, Illinois. Unusual, but enjoyable. Could be a fine fireside sip given the smokiness.

My Score

Notes: ______________________________

____

____

Appearance (1-5) ____ Nose (1-25) ____
Taste (1-25) ____ Finish (1-25) ____
Complexity (1-10) ____ Overall Impression (1-10) ____

SCORE ____

Corner Creek 10-Year-Old

Proof:	88
Age:	10 years
Type:	Straight
Style:	Non-Chill Filtered, Small Batch
Mash Bill:	Not available
Color:	Dark Straw
Price:	$$$

Mike

Nose: Vanilla, oak, and a hint of cherries or dates.
Taste: Vanilla, oak, and a little pepper and dates.
Finish: Medium long and peppery.

Susan

Nose: Ripe apples, light caramel, and some cereal.
Taste: Caramel, pepper, lingering dark fruit.
Finish: Long and fruity with some pepper and oak.

Notes

Drinks hotter than it's 88 proof. Ice cube mellows it. Kentucky Artisan Distillery, Crestwood, Kentucky, https://kentuckyartisandistillery.com.

My Score

Notes: ______________________________

____

____

Appearance (1-5) ____ Nose (1-25) ____
Taste (1-25) ____ Finish (1-25) ____
Complexity (1-10) ____ Overall Impression (1-10) ____

SCORE ____

Corner Creek Reserve Bourbon Whiskey

Proof: 88
Age: NAS
Type: Reserve Bourbon Whiskey
Style: Single Barrel
Mash Bill: Proprietary
Color: Light Amber
Price: $$

Mike

Nose: Caramel and vanilla with sweet corn and spice. A little oak in the background.
Taste: Very light. Vanilla and pepper spice. Not complicated. No burn. But not much flavor.
Finish: This is the saving grace. Nice peppery finish with oak tannins lasting a long time.

Susan

Nose: Candied fruit, candied almonds, a little caramel and vanilla.
Taste: Buttered popcorn, crème brûlée, and fruit cocktail. Luscious mouthfeel.
Finish: Long and smooth. Sweet finish dries to a little oaky spice.

Notes

Water helps bring out the caramel notes in Corner Creek.

My Score

Notes: ________________________________

____

____

Appearance (1-5) ____ Nose (1-25) ____
Taste (1-25) ____ Finish (1-25) ____
Complexity (1-10) ____ Overall Impression (1-10) ____
SCORE ____

Cream of Kentucky

Proof: 102
Age: Minimum 11 ½ years
Type: Straight
Style: Extra Aged
Mash Bill: Not available
Color: Light Amber
Price: $$$$

Mike

Nose: Vanilla and marshmallows with a little banana.
Taste: Vanilla, nutmeg, and allspice with some banana, fruit, and oak.
Finish: Surprisingly short with some oak.

Susan

Nose: Bananas and ripe apples with some pear followed by waves of vanilla and a little cinnamon.
Taste: Very ripe fruit with vanilla, caramel, and some baking spice.
Finish: Spice and oak that lingers on the tongue.

Notes

A little water enriches the mouthfeel. Sourced and selected by Jim Rutledge, former master distiller at Four Roses. Revival of a historic premium brand. Proof and age vary with releases.

My Score

Notes: ______________________________

____

____

Appearance (1-5) ____ Nose (1-25) ____
Taste (1-25) ____ Finish (1-25) ____
Complexity (1-10) ____ Overall Impression (1-10) ____
SCORE ____

Crooked Furrow

Proof: 92
Age: Minimum 4 years
Type: Straight
Style: Traditional
Mash Bill: Not available
Color: Straw
Price: $$$

Mike

Nose: Vanilla and corn with a hint of oak.
Taste: Vanilla, corn, pepper, oak.
Finish: Dry with oak and white pepper.

Susan

Nose: Caramel, light herbs, cinnamon, perhaps some cherries.
Taste: Caramel, oak, and peppery spice.
Finish: Pleasantly peppery with some caramel undertones and a touch of tannins.

Notes

Grains sourced from the Red River Valley. North Dakota's first bourbon.

My Score

Notes: ____________________________________

____

____

Appearance (1-5) ____ Nose (1-25) ____
Taste (1-25) ____ Finish (1-25) ____
Complexity (1-10) ____ Overall Impression (1-10) ____
SCORE ____

C.W. Irwin

Proof: 80
Age: NAS
Type: Straight
Style: Four Grain, Craft
Mash Bill: 65% corn, 18% rye, 9% barley, 8% wheat*
Color: Very Pale Straw
Price: $$

Mike

Nose: Very light – vanilla and bread dough.
Taste: Brown vodka, Very light. Some vanilla and corn.
Finish: Long and dry with oak.

Susan

Nose: Light, some fruit and yeasty dough.
Taste: Light. Butterscotch, but little else.
Finish: Begins smoothly, but a light woody aftertaste.

Notes

From Oregon Spirit Distiller, Bend, Oregon. The mash bill recipe* from information received from the purchaser of the bourbon. Named in honor of Craig Weber Irwin, brother of Brad Irwin. Brad and his wife Kathy are the distillers and made the first two barrels in their garage. Craig bought the barrels, thus funding their new distilling enterprise.

My Score

Notes: ______________________________
____
____

Appearance (1-5) ____ Nose (1-25) ____
Taste (1-25) ____ Finish (1-25) ____
Complexity (1-10) ____ Overall Impression (1-10) ____
SCORE ____

Cyrus Noble

Proof: 90
Age: NAS
Type: Straight
Style: Small Batch
Mash Bill: Not available
Color: Straw
Price: $$

Mike

Nose: Corn and vanilla. Hint of chocolate.
Taste: Sweet vanilla and berries. Hint of oak.
Finish: Short and sweet. Hint of oak.

Susan

Nose: Very light caramel. Not complex.
Taste: Corn and black pepper with underlying caramel and fruit.
Finish: Short.

Notes

Sourced from an unnamed Kentucky distillery and bottled by Haas Brothers, San Francisco, California.

My Score

Notes: ____

____

____

Appearance (1-5) ____ Nose (1-25) ____
Taste (1-25) ____ Finish (1-25) ____
Complexity (1-10) ____ Overall Impression (1-10) ____

SCORE ____

Dancing Pines Bourbon

Proof:	88
Age:	Less than 4 years
Type:	Not Straight
Style:	Small Batch
Mash Bill:	Not available
Color:	Light Straw
Price:	$$$

Mike

Nose: Corn and vanilla with a hint of apricots.
Taste: Corn and vanilla with apricots and oak.
Finish: Starts fruity and dries to oak.

Susan

Nose: Light caramel, white pepper, cinnamon.
Taste: Caramel, some vanilla, oak, sweet fruit.
Finish: Short and dry.

Notes

Made by Dancing Pines Distillery, Loveland, Colorado. This was small batch #17.

My Score

Notes: ____________________________________

____

____

Appearance (1-5) ____ Nose (1-25) ____
Taste (1-25) ____ Finish (1-25) ____
Complexity (1-10) ____ Overall Impression (1-10) ____

SCORE ____

David James Bourbon

Proof: 111.2
Age: 8 years, 10 months
Type: Straight
Style: Barrel Proof, Small Batch
Mash Bill: Not available
Color: Light Straw
Price: $$$

Mike

Nose: Caramel, apples, and a little leather and oak.
Taste: Caramel, apples with a little vanilla and nutmeg. A hint of oak.
Finish: Long and dry with lots of oak and a bit of pepper.

Susan

Nose: Caramel, a little apricot, and brown sugar.
Taste: Caramelized bananas, crème brulee, and cinnamon.
Finish: Sweet at the beginning and quickly drying to peppery oak.

Notes

Sourced from Tennessee. Bottled by Neeley Family Distillery, Sparta, Kentucky. Batch of 25 barrels.

My Score

Notes: ____________________________________

____

____

Appearance (1-5) ____ Nose (1-25) ____
Taste (1-25) ____ Finish (1-25) ____
Complexity (1-10) ____ Overall Impression (1-10) ____
SCORE ____

Deerhammer Four Grain Bourbon

Proof: 88
Age: 2 years
Type: Straight
Style: Four Grain, Pot Distilled
Mash Bill: 65% corn, 20% red winter wheat, 10% oats, 5% roasted barley
Color: Straw
Price: $$

Mike

Nose: Corn, vanilla, a little apricot fruit.
Taste: Caramel, apricots, white pepper, and oak.
Finish: Medium long with a nice spice finish.

Susan

Nose: Multigrain cereal with vanilla and green apple.
Taste: Faintly smoky with some multigrain flavors, faint herb, and apple peel.
Finish: Surprisingly sweet before drying to light oak.

Notes

Deerhammer Distilling Company, Buena Vista, Colorado. Fermented with open-air yeast. Uses grain grown in Colorado. Limited release expressions include those finished in ex-Belgian beer and port wine barrels.

My Score

Notes: ______________________________

____

____

Appearance (1-5) ____ Nose (1-25) ____
Taste (1-25) ____ Finish (1-25) ____
Complexity (1-10) ____ Overall Impression (1-10) ____

SCORE ____

Delaware Phoenix

Proof: 100
Age: Under 12 months
Type: Not Straight
Style: Craft
Mash Bill: ~ 69% corn, 18% rye, 13% malted barley
Color: Amber
Price: $$

Mike

Nose: Honey and vanilla with corn and a bit of yeasty bread dough.
Taste: Corn and vanilla with some pepper spice. Fine tobacco and oak.
Finish: Long. Starts spicy but gets drier with oak tannins.

Susan

Nose: Corn! No mistaking that this is a corn-based whiskey. Caramel emerges with time in the glass.
Taste: Vanilla wafers, nutmeg, cinnamon. Creamy mouthfeel.
Finish: Long and dry. Very smooth.

Notes

Pot-stilled, craft bourbon from Walton, New York distiller Cheryl Lin is aged in small oak casks for less than a year. Impressive character for such a young bourbon. Note that the pricing is for 375ml bottles. It is not bottled in fifths.

My Score

Notes: ____________________________________
____
____

Appearance (1-5) ____ Nose (1-25) ____
Taste (1-25) ____ Finish (1-25) ____
Complexity (1-10) ____ Overall Impression (1-10) ____
SCORE ____

Dettling 1867 Bottled-in-Bond

Proof: 100
Age: 4 years
Type: Straight
Style: Bonded, Six Grain
Mash Bill: Undisclosed proportions of corn, rye, oats, roasted malt wheat, roasted rye malt, roasted barley malt
Color: Amber
Price: $$

Mike

Nose: Caramel, apples, and pears.
Taste: Caramel, apples, cardamom, and allspice with a little oak.
Finish: Long with oak and spice.

Susan

Nose: Dark chocolate-coated caramel, a whiff of roasted coffee beans, dark fruit.
Taste: Cinnamon candy, ripe apple, chocolate/caramel/coffee with new leather. Complex.
Finish: Flavors fade into sweet oak.

Notes

Balanced. No one flavor predominates. A limited 2019 release.

My Score

Notes: ______________________________

____

____

Appearance (1-5) ____ Nose (1-25) ____
Taste (1-25) ____ Finish (1-25) ____
Complexity (1-10) ____ Overall Impression (1-10) ____

SCORE ____

Dettling 1867 Cask Strength

Proof: 107.8
Age: 35 months
Type: Straight
Style: Cask Strength, Single Barrel, Finished, Six Grain
Mash Bill: Undisclosed proportions of corn, rye, oats, roasted malt wheat, roasted rye malt, roasted barley malt
Color: Dark Amber
Price: $$

Mike

Nose: Brown sugar, caramel, hint of chocolate and a little corn. Apples in the background.
Taste: Brown sugar, apples, pepper, and oak.
Finish: Long. Starts sweet, but dries out with oak tannins.

Susan

Nose: Burnt caramel, dates, figs, and toasted marshmallows.
Taste: Figs, buttery caramel, and milk chocolate.
Finish: Sweet caramel dries to baking spice.

Notes

Finished in a second, charred oak barrel. Water releases more chocolate. Check website for more expressions.

My Score

Notes: ______________________________

____

____

Appearance (1-5) ____ Nose (1-25) ____
Taste (1-25) ____ Finish (1-25) ____
Complexity (1-10) ____ Overall Impression (1-10) ____

SCORE ____

District Made Straight Bourbon Whiskey

Proof: 95
Age: NAS
Type: Straight
Style: Four Grain
Mash Bill: Uses Hickory King corn, wheat, malted rye, malted barley
Color: Straw
Price: $$

Mike

Nose: Bread dough, vanilla, apricots.
Taste: Vanilla, corn, apricot, and a hint of pepper.
Finish: Short and dry with oak and pepper.

Susan

Nose: Corn, vanilla, apricots.
Taste: Frosted corn flakes, vanilla, ripe apples, and some new leather.
Finish: Dry and spicy.

Notes

There is also a limited release bottled-in-bond.

My Score

Notes: ____________________________________

____

____

Appearance (1-5) ____ Nose (1-25) ____
Taste (1-25) ____ Finish (1-25) ____
Complexity (1-10) ____ Overall Impression (1-10) ____

SCORE ____

Driftless Glen

Proof:	96
Age:	4 years
Type:	Straight
Style:	Small Batch, Non-Chill Filtered
Mash Bill:	60% corn, 20% rye, 20 % malted barley
Color:	Straw
Price:	$$

Mike

Nose: Vanilla, apricot, pepper, oak.
Taste: Vanilla, apricots, white pepper, oak.
Finish: Long with oak and pepper.

Susan

Nose: Caramel, cocoa, little fruit or spice.
Taste: Caramel, orange peel, cinnamon, brown sugar.
Finish: Medium long and oaky.

Notes

Well balanced. Driftless Glen Distillery, Baraboo, Wisconsin.

My Score

Notes: ______________________________

____

____

Appearance (1-5) ____ Nose (1-25) ____
Taste (1-25) ____ Finish (1-25) ____
Complexity (1-10) ____ Overall Impression (1-10) ____
SCORE ____

Dry Fly Bourbon 101

Proof: 101
Age: NAS
Type: Straight
Style: Small Batch
Mash Bill: Not available
Color: Straw
Price: $$$

Mike

Nose: Corn, bread dough, and vanilla.
Taste: Corn and vanilla with a hint of pepper.
Finish: Dry oak and corn. Gets sweet with some honey and oak.

Susan

Nose: Grilled corn and herbs. Succotash!
Taste: Very vegetative. Some vanilla sweetness.
Finish: Dry and oaky. A hint of vanilla at the end.

Notes

From Dry Fly Distilling, Spokane, Washington. Sold in 375ml bottles. This was small batch #2.

My Score

Notes: ______________________________

____

____

Appearance (1-5) ____ Nose (1-25) ____
Taste (1-25) ____ Finish (1-25) ____
Complexity (1-10) ____ Overall Impression (1-10) ____

SCORE ____

Duke

Proof: 88
Age: Batched election of 5-10 year old barrels
Type: Straight
Style: Traditional
Mash Bill: High Rye
Color: Dark Straw
Price: $$

Mike

Nose: Corn and saddle leather. Hint of vanilla oak. Not complex.
Taste: Thin and watery mouthfeel – caramel and some pepper spice.
Finish: Dry and peppery. Not bad, but a bit hot.

Susan

Nose: Caramel, spiced pecans. Ripe apples and emerge as it sits.
Taste: Caramel apples evolving into hot Corn with pepper.
Finish: Long and peppery hot.

Notes

Could make a good Manhattan with a strong, sweet vermouth. Bottle says "Monument Valley Distillers" and "Duke Spirits, Lawrenceburg, Kentucky." It is sourced from Wild Turkey.

My Score

Notes: ______________________________

____

____

Appearance (1-5) ___ Nose (1-25) ___
Taste (1-25) ___ Finish (1-25) ___
Complexity (1-10) ___ Overall Impression (1-10) ___

SCORE ___

Eighteen 33

Proof: 80.8
Age: 10 years
Type: Straight
Style: Extra Aged
Mash Bill: Not available
Color: Straw
Price: $$$

Mike

Nose: Very light. Vanilla with a hint of fruit.
Taste: Better than the nose indicates. Fruit and spice with caramel.
Finish: Starts sweet with caramel and nicely dries to oak.

Susan

Nose: Caramel corn, orange blossom, vanilla sugar.
Taste: Cinnamon, some chocolate, lots of caramel.
Finish: Warm and spicy.

Notes

Very nice sip bottled by Boone County Distillery, Boone, County, Kentucky. Sourced from MGP in Lawrenceburg, Indiana.

My Score

Notes: ______________________________

____

____

Appearance (1-5) ____ Nose (1-25) ____
Taste (1-25) ____ Finish (1-25) ____
Complexity (1-10) ____ Overall Impression (1-10) ____
SCORE ____

Featherbone Bourbon

Proof: 90
Age: NAS
Type: Not Straight
Style: Small Batch
Mash Bill: Not available
Color: Dark Straw
Price: $$$

Mike

Nose: Bread dough and vanilla with a hint of nutmeg.
Taste: Corn, vanilla and lots of oak.
Finish: Dry with lots of oak.

Susan

Nose: Pine resin, pears, underlying oak.
Taste: Sweet oak, some roasted corn, new leather.
Finish: Oaky, peppery.

Notes

Made by Journeyman Distillery, Three Oaks, Michigan. This was from small batch PB3, Bottle 17.

My Score

Notes: ______________________________

____

____

Appearance (1-5) ___ Nose (1-25) ___
Taste (1-25) ___ Finish (1-25) ___
Complexity (1-10) ___ Overall Impression (1-10) ___
SCORE ___

Fern Creek Single Barrel

Proof:	113.3
Age:	5 years
Type:	Straight
Style:	Single Barrel
Mash Bill:	78% corn, 10% rye, 12% malted barley
Color:	Light Straw
Price:	$$$

Mike

Nose: French vanilla cream, some baking spice and oak.

Taste: Vanilla, with a little pear and apple, allspice, and oak.

Finish: Short to medium ling with oak, spice, and lingering vanilla.

Susan

Nose: Brown sugar, crème brulee, light cinnamon and a little fruit.

Taste: Sugar and spice. Some plums and cinnamon.

Finish: Short and spicy.

Notes

Sourced bourbon that also has a small batch release.

My Score

Notes: ______________________________

____

____

Appearance (1-5)	____	Nose (1-25)	____
Taste (1-25)	____	Finish (1-25)	____
Complexity (1-10)	____	Overall Impression (1-10)	____

SCORE ____

F•E•W Bourbon

Proof: 93
Age: NAS
Type: Straight
Style: Traditional
Mash Bill: 70% corn, 20% Northern rye, 10% 2-row barley
Color: Dark Straw
Price: $$$

Mike

Nose: Corn vanilla, and oak.
Taste: Dark fruit, some cinnamon, allspice, and vanilla.
Finish: Long and spicy with oak and some lingering apple.

Susan

Nose: Caramel and some candied plums.
Taste: Carmel, vanilla, candied ginger, with dark fruit.
Finish: Long with pepper and ginger.

Notes

Look for the 10th Anniversary expression, too.

My Score

Notes: ______________________________________
____
____

Appearance (1-5) ____ Nose (1-25) ____
Taste (1-25) ____ Finish (1-25) ____
Complexity (1-10) ____ Overall Impression (1-10) ____

SCORE ____

F•E•W Bourbon Bottled-in-Bond

Proof: 100
Age: At least 4 years
Type: Straight
Style: Bonded
Mash Bill: 70% corn, 20% Northern rye, 10% 2-row barley
Color: Amber
Price: $$$

Mike

Nose: Dates and plums with a hint of spice and oak.
Taste: Lots of prunes and dates with cinnamon, allspice, vanilla, and oak.
Finish: Long with oak, spices, and lingering dates.

Susan

Nose: Peaches, pears, dark cherry, allspice, and vanilla.
Taste: Crème brulee with some cinnamon, a dash of cardamom, milk chocolate and baked apples.
Finish: Long and smooth with light oak and not trace of bitterness.

Notes

Balanced and complex.

My Score

Notes: ____________________________________

____

____

Appearance (1-5) ____ Nose (1-25) ____
Taste (1-25) ____ Finish (1-25) ____
Complexity (1-10) ____ Overall Impression (1-10) ____

SCORE ____

Filibuster Dual Cask

Proof: 90
Age: NAS
Type: Straight
Style: Special Finish
Mash Bill: 75% corn, 21% rye, 4% malted barley
Color: Light Amber
Price: $$

Mike

Nose: Caramel and vanilla with some apple and spice.
Taste: Corn and vanilla with nutmeg.
Finish: Dry with oak and spice.

Susan

Nose: Caramel, sweet spice, light oak and leather.
Taste: Ripe apples, bit of barnyard (in a good way) and sweet oak.
Finish: Dries to peppery oak.

Notes

From Filibuster Distillery, Maụretown, Virginia. A "master blender" choses whiskies for finishing in French oak after aging in charred oak. This was Small batch #2, bottle #866.

My Score

Notes: ______________________________

____

____

Appearance (1-5) ____ Nose (1-25) ____
Taste (1-25) ____ Finish (1-25) ____
Complexity (1-10) ____ Overall Impression (1-10) ____

SCORE ____

Fireside Colorado Bourbon

Proof:	80
Age:	At least 2 years
Type:	Straight
Style:	Pot Distilled, Sweet Mash
Mash Bill:	70% corn, 20% rye, 10% malted barley
Color:	Pale Straw
Price:	$$

Mike

Nose: New mown lawn with a hint of vanilla and apricots.
Taste: Thin mouthfeel. Vanilla and apricots with a hint of spice.
Finish: Short and slightly dry with oak.

Susan

Nose: Spicy with cardamom and some citrus.
Taste: Predominantly spices with some vanilla.
Finish: Very dry and spicy.

Notes

The distillery also has expressions with a little more age and proof: bottled-in-bond and barrel proof.

My Score

Notes: ____

____

____

Appearance (1-5) ____ Nose (1-25) ____
Taste (1-25) ____ Finish (1-25) ____
Complexity (1-10) ____ Overall Impression (1-10) ____
SCORE ____

Flag Hill Straight Bourbon

Proof: 90
Age: 2 years
Type: Straight
Style: Small Batch
Mash Bill: Not available
Color: Pale Straw
Price: $$$

Mike

Nose: Corn, vanilla, yeast, and a hint of banana.
Taste: Vanilla, corn, pepper, banana.
Finish: Short and medicinal.

Susan

Nose: Corn husk, corn dough, pepper.
Taste: Some sweetness arrives on the palate, but very grain forward.
Finish: Short and dry.

Notes

Very young. We hope the distillery releases some older expressions in the future and look forward to trying it.

My Score

Notes: ______________________________

____

____

Appearance (1-5) ____ Nose (1-25) ____
Taste (1-25) ____ Finish (1-25) ____
Complexity (1-10) ____ Overall Impression (1-10) ____

SCORE ____

Flatboat

Proof: 90
Age: NAS
Type: Straight
Style: Traditional
Mash Bill: Not available
Color: Light Straw
Price: $$

Mike

Nose: Honeysuckle and bananas with a hint of spice.
Taste: Sweet honey and apples with baking spices.
Finish: Long. Starts sweet and dries nicely with some oak.

Susan

Nose: Bananas and some vanilla.
Taste: Banana carries through to the palate with some light caramel and fruit.
Finish: Hot on the tongue.

Notes

Bottled by The Founders Company, Louisville, Kentucky. Sourced from Barton, Bardstown, Kentucky.

My Score

Notes: ______________________________

____

____

Appearance (1-5) ____ Nose (1-25) ____
Taste (1-25) ____ Finish (1-25) ____
Complexity (1-10) ____ Overall Impression (1-10) ____

SCORE ____

Frey Ranch Bourbon

Proof:	90
Age:	NAS
Type:	Straight
Style:	Four Grain, Non-Chill Filtered
Mash Bill:	Corn, winter rye, winter wheat, 2-row barley
Color:	Dark Straw
Price:	$$$

Mike

Nose: Caramel and spice with oak and tobacco.
Taste: Caramel, apples, allspice, nutmeg, and a little oak.
Finish: Long with oak and spice.

Susan

Nose: Melons, vanilla, nutmeg. Combination of fruit stand and flower shop.
Taste: Fruit explodes on the tongue, but well-balanced by a good dash of peppery spice.
Finish: Long in the mouth and lingers after that in the throat.

Notes

Very smooth. All grains grown on the Frey Hill ranch, which has been farmed since 1854. Frey Ranch Distillery, Fallon, Nevada, https://freyranch.com.

My Score

Notes: ______________________________

____

____

Appearance (1-5) ____ Nose (1-25) ____
Taste (1-25) ____ Finish (1-25) ____
Complexity (1-10) ____ Overall Impression (1-10) ____

SCORE ____

Garrison Brothers

Proof: 94
Age: 4 years
Type: Straight
Style: Craft, Wheated
Mash Bill: Not available
Color: Light Amber
Price: $$$

Mike

Nose: Sweet corn and yeast. Cornbread dough. Not complex. Just a hint of oak.

Taste: Sweet corn and spice. Bit of pepper but rotten peach as well.

Finish: Long and sickly sweet – rotten peaches. Dries to oak.

Susan

Nose: Caramel, spiced pecans. Ripe apples emerge as it sits.

Taste: Caramel apples evolving into hot has some pepper.

Finish: Long and peppery hot.

Notes

Garrison Brothers, Hye, Texas. Aged in small casks rather than 53 gallon barrels. Uses locally sourced white corn and winter wheat. This was bottle No. 14882.

My Score

Notes: ______________________________

____

____

Appearance (1-5)	___	Nose (1-25)	___
Taste (1-25)	___	Finish (1-25)	___
Complexity (1-10)	___	Overall Impression (1-10)	___

SCORE ___

Garrison Brothers Small Batch Texas Straight

Proof: 94
Age: NAS
Type: Straight
Style: Small Batch, Sweet Mash, Wheated
Mash Bill: 74% corn, 15% red winter wheat, 11% malted barley
Color: Dark Straw
Price: $$$

Mike

Nose: Bread dough and vanilla. Not much else.
Taste: Bread dough, vanilla, a bit of apple and oak.
Finish: Best aspect. Nice fruit notes, with pepper and oak.

Susan

Nose: Roasted corn and vanilla.
Taste: Buttered corn, ripe apples, and almonds.
Finish: Smooth and nutty.

Notes

Other expressions include a single barrel, barrel proof (Cowboy Bourbon), double-barreled (Balmorhea), and one made with rye.

My Score

Notes: ______________________________
____
____

Appearance (1-5) ____ Nose (1-25) ____
Taste (1-25) ____ Finish (1-25) ____
Complexity (1-10) ____ Overall Impression (1-10) ____
SCORE ____

George Remus

Proof: 90.7
Age: NAS
Type: Not Straight
Style: Small Batch
Mash Bill: Not available
Color: Light Straw
Price: $$

Mike

Nose: Very young. Lots of fusel oil. Some corn vanilla.
Taste: Sweet corn and oak tannins. Corn with some pepper spice. Tastes young.
Finish: Astringent and dry with oak tannins.

Susan

Nose: Mineral oil. Needs more time in the barrel.
Taste: Corn, a little oak. Not much else.
Finish: Starts sweet. Ends hot.

Notes

Batch 1. Queen City Whiskey, Bardstown, Kentucky. But it is sourced from MGP in Lawrenceburg, Indiana. Named for the famous Cincinnati bootlegger.

My Score

Notes: ______________________________

____

____

Appearance (1-5) ____ Nose (1-25) ____
Taste (1-25) ____ Finish (1-25) ____
Complexity (1-10) ____ Overall Impression (1-10) ____

SCORE ____

George Remus Straight

Proof: 94
Age: NAS, but "over 4 years" according to distillery
Type: Straight Bourbon
Style: Traditional
Mash Bill: Not available
Color: Straw
Price: $$

Mike

Nose: Berries and vanilla with some spice.
Taste: Caramel and berries with nutmeg and oak.
Finish: Sweet notes with a little oak.

Susan

Nose: Berries, caramel, and baking spices.
Taste: Caramel and nuts, Some underlying astringency.
Finish: Black pepper with a surprising sweet note at the end.

Notes

Brand acquired by contract distilling giant Midwest Grain Producers of Lawrenceburg, Indiana. Named after an infamous Cincinnati bootlegger.

My Score

Notes: ______________________________

____

____

Appearance (1-5) ___ Nose (1-25) ___
Taste (1-25) ___ Finish (1-25) ___
Complexity (1-10) ___ Overall Impression (1-10) ___
SCORE ___

Glenns Creek Swheat

Proof: 105
Age: 56 months
Type: Straight
Style: Wheated, Single Barrel, Non-Chill Filtered, Pot Distilled
Mash Bill: 87% corn, 10% rye, 3% malted barley
Color: Dark Straw
Price: $$$

Mike

Nose: Vanilla, prunes, hint of chocolate, oak.
Taste: Vanilla, apricots, cinnamon, nutmeg, and oak.
Finish: Medium long with oak and cinnamon.

Susan

Nose: Spicy oak, some peach peel, a touch of cocoa.
Taste: Cinnamon sugar, light cocoa, sweet oak.
Finish: Medium long with lingering cinnamon and caramel.

Notes

Glenns Creek Distillery, Frankfort, Kentucky. Barrel #18. Located in the former Old Crow Distillery.

My Score

Notes: ____________________

____

____

Appearance (1-5) ____ Nose (1-25) ____
Taste (1-25) ____ Finish (1-25) ____
Complexity (1-10) ____ Overall Impression (1-10) ____

SCORE ____

Grand River Baby Bourbon

Proof: 80.9
Age: 6 months or less
Type: Not Straight
Style: Small Batch, Single Barrell
Mash Bill: Not available
Color: Very, Very Light Straw
Price: $$$

Mike

Nose: Honey and citrus with a hint of yeast.
Taste: Citrus and vanilla with some spice.
Finish: Short with a little spice.

Susan

Nose: Sweet corn, a little new leather.
Taste: Frosted corn flakes. Buttery mouthfeel.
Finish: Very sweet with a dash of pepper at the end.

Notes

Made by Grand River Spirits, Carbondale, Illinois. This was from small batch #8, single barrel #23. Aged in 15 or 30 gallon white oak barrels.

My Score

Notes: ______________________________

____

____

Appearance (1-5) ____ Nose (1-25) ____
Taste (1-25) ____ Finish (1-25) ____
Complexity (1-10) ____ Overall Impression (1-10) ____
SCORE ____

Grass Widow

Proof: 91
Age: NAS
Type: Straight
Style: High Rye
Mash Bill: 60% corn, 36% rye, 4% malted barley
Color: Straw
Price: $$$

Mike

Nose: Caramel corn with a little fruit and oak.
Taste: Caramel apples and pepper with a hint of oak.
Finish: Long and dry with oak and a hint of spice.

Susan

Nose: Carmel corn with apples, hazelnuts, a touch of spice.
Taste: Vanilla, apples, cinnamon, a bit of citrus.
Finish: Sweet before takings its time to dry into oak.

Notes

Two James Spirits Company, Detroit and Grand Rapids, Michigan Part of a portfolio of spirits that includes gin, rye, and malt whiskeys. Original art adores the walls of the distillery tasting room and changes quarterly.

My Score

Notes: ____________________

____

____

Appearance (1-5) ____ Nose (1-25) ____
Taste (1-25) ____ Finish (1-25) ____
Complexity (1-10) ____ Overall Impression (1-10) ____

SCORE ____

Gun Fighter American Bourbon Whiskey Double Cask

Proof: 100
Age: 6 months
Type: Straight
Style: Finished
Mash Bill: Not available
Color: Dark Straw
Price: $$

Mike

Nose: Vanilla and oak with a hint of pears and spice.
Taste: Vanilla, and pears with nutmeg and cinnamon.
Finish: Long and dry with oak and spice.

Susan

Nose: Fruity, with oak and spice.
Taste: Baked apple, dates, vanilla, and baking spice.
Finish: Warm and peppery.

Notes

Ages 6 months in American white oak and then transferred to ex-port barrel of French oak.

My Score

Notes: ____________________

____

____

Appearance (1-5) ____ Nose (1-25) ____
Taste (1-25) ____ Finish (1-25) ____
Complexity (1-10) ____ Overall Impression (1-10) ____
SCORE ____

Hard Truth Bottled-in-Bond Four Grain

Proof: 100
Age: 5 years
Type: Straight
Style: Bonded, Small Batch, Sweet Mash
Mash Bill: 78% corn, 9% rye, 9% wheat 4% malted barley
Color: Straw
Price: $$$

Mike

Nose: Vanilla, apricots, peaches, baking spice, oak.
Taste: Vanilla, peaches, cinnamon, oak.
Finish: Long with oak and cinnamon.

Susan

Nose: Rich toffee, peaches, some milk chocolate, tobacco leaf.
Taste: Fruit pops out first. Ripe apples with cinnamon and caramel.
Finish: Long with baking spices.

Notes

Hard Truth Distilling Co., Nashville, Indiana. Velvety mouthfeel.

My Score

Notes: ____________________

____

____

Appearance (1-5) ____ Nose (1-25) ____
Taste (1-25) ____ Finish (1-25) ____
Complexity (1-10) ____ Overall Impression (1-10) ____

SCORE ____

Hard Truth Bottled-in-Bond Sweet Mash Wheated

Proof: 100
Age: 5 years
Type: Straight
Style: Bonded, Small Batch, Sweet Mash, Wheated
Mash Bill: 69% corn, 19% wheat 12% malted barley
Color: Dark Straw
Price: $$$

Mike

Nose: Vanilla, apricots, pepper, and oak.
Taste: Vanilla, apricots, black pepper, hazelnuts, oak.
Finish: Long with oak and pepper.

Susan

Nose: Vanilla, unripe berries, light cinnamon and oak.
Taste: Cinnamon sugar, nutmeg, vanilla frosting, cashews, light apple.
Finish: Long with peppery oak.

Notes

Hard Truth Distilling Co., Nashville, Indiana. Fine example of a wheated expression.

My Score

Notes: ____________________________________

____

____

Appearance (1-5) ____ Nose (1-25) ____
Taste (1-25) ____ Finish (1-25) ____
Complexity (1-10) ____ Overall Impression (1-10) ____

SCORE ____

Hard Truth Sweet Mash

Proof:	90
Age:	NAS
Type:	Straight
Style:	Small Batch, Sweet Mash
Mash Bill:	73% corn, 19% rye, 8% malted barley
Color:	Straw
Price:	$$

Mike

Nose: Vanilla, apples, allspice, oak.
Taste: Vanilla, apples, allspice, tobacco, and oak.
Finish: Long with oak, spice, and tobacco.

Susan

Nose: Brown sugar, a whiff of milk chocolate, and some citrus.
Taste: Vanilla, candied ginger, light orange zest.
Finish: Medium long and tingly.

Notes

Hard Truth Distilling Co., Nashville, Indiana. Balanced with a rich mouthfeel.

My Score

Notes: ______________________________

____

____

Appearance (1-5) ____ Nose (1-25) ____
Taste (1-25) ____ Finish (1-25) ____
Complexity (1-10) ____ Overall Impression (1-10) ____
SCORE ____

Hartfield & Co.

Proof: 100
Age: NAS
Type: Not Straight
Style: Small Batch
Mash Bill: 62% corn, 19% rye, 19% malted barley
Color: Straw
Price: $$$

Mike

Nose: Berries and corn with vanilla, honey, and tobacco.
Taste: Berries, corn, vanilla and pepper.
Finish: Dry oak and pepper with a hint of fruit.

Susan

Nose: Corn and rye. Whiff of honey.
Taste: Roasted corn, leather, dried apples.
Finish: Dry new oak. Touch of woodsmoke.

Notes

Made by Hartfield Distillery, Paris, Kentucky. This was from small batch #10, bottle 208 of 296. Hartfield is the first distillery to open in Bourbon County, Kentucky since Prohibition.

My Score

Notes: ____________________________________
____
____

Appearance (1-5) ____ Nose (1-25) ____
Taste (1-25) ____ Finish (1-25) ____
Complexity (1-10) ____ Overall Impression (1-10) ____

SCORE ____

Henry DuYore's

Proof: 91.3
Age: NAS
Type: Straight
Style: Small Batch
Mash Bill: 56% corn, 31% rye, 13% malted barley
Color: Dark Straw
Price: $$$

Mike

Nose: Caramel with a hint of turpentine and ripe fruit.
Taste: Caramel and apples with oak. Better than the nose would indicate.
Finish: Sweet with caramel, hint of chocolate and oak.

Susan

Nose: Vanilla, ripe apples, new leather.
Taste: Caramel, sweet oak, raisins.
Finish: Long and dry with peppery ending and a little cocoa powder.

Notes

Made by Ransom Spirits, Sheridan, Oregon. This was from small batch #13, bottle #1354. Label depicts a bootlegger. "Henry DuYore is not my real name."

My Score

Notes: ____

____

____

Appearance (1-5) ____ Nose (1-25) ____
Taste (1-25) ____ Finish (1-25) ____
Complexity (1-10) ____ Overall Impression (1-10) ____

SCORE ____

High N' Wicked

Proof: 104
Age: 5 years
Type: Straight
Style: Barrel Strength
Mash Bill: 51% corn, 39% rye, 10% malted barley
Color: Straw
Price: $$$

Mike

Nose: Fruit forward. Apricots, peaches, vanilla, and oak.
Taste: Vanilla, apricots, black pepper, oak.
Finish: Long with oak and lots of black pepper.

Susan

Nose: Vanilla, horehound, unripe peach.
Taste: Nutty vanilla, cinnamon sugar, apple.
Finish: Dry but not bitterly tannic.

Notes

Distilled at Kentucky Artisan Distillery, Crestwood, Kentucky. Brand founded in 2019 by two former Brown-Forman senior executives, W.L. Lyons Brown, III, and Kevin E. Sachs.

My Score

Notes: ____

____

____

Appearance (1-5) ____ Nose (1-25) ____
Taste (1-25) ____ Finish (1-25) ____
Complexity (1-10) ____ Overall Impression (1-10) ____

SCORE ____

High West Whiskey

Proof: 92
Age: NAS
Type: Straight Bourbon
Style: Small Batch, Non-Chill Filtered
Mash Bill: Not available
Color: Dark Straw
Price: $$

Mike

Nose: Honey and vanilla with corn. A little sweet spice.
Taste: A bit thin in the mouth. Vanilla and corn with some sweet clove.
Finish: Long and sweet with the spices – clove and nutmeg.

Susan

Nose: Prominent vanilla with touch of sweet spices faint floral note.
Taste: Vanilla and spices with a very faint note of green apple.
Finish: Medium long, warm and spicy.

Notes

Made by High West Distillery, Park City, Utah. Boasts it is the only ski in gastro-distillery and we believe them.

My Score

Notes: ____

____

____

Appearance (1-5) ____ Nose (1-25) ____
Taste (1-25) ____ Finish (1-25) ____
Complexity (1-10) ____ Overall Impression (1-10) ____

SCORE ____

High West Whiskey American Prairie

Proof: 92
Age: 2 years
Type: Straight
Style: Small Batch
Mash Bill: Not available
Color: Straw
Price: $$

Mike

Nose: Apricots and honey with a hint of spice.
Taste: Fruity with apricots and apples. Some corn vanilla and spice.
Finish: Very nice with oak and spice. A hint of fruit sweetness.

Susan

Nose: Corn, fresh evergreens.
Taste: Brown sugar, apples, sweet spices.
Finish: Dry and peppery with touch of sweetness.

Notes

This is a mingling of straight bourbons. High West Distillery is in Park City, Utah. This bottle was from batch #16C15. The distillery donates 10% of profits to prairie preservation.

My Score

Notes: ________________________________

____

____

Appearance (1-5) ____ Nose (1-25) ____
Taste (1-25) ____ Finish (1-25) ____
Complexity (1-10) ____ Overall Impression (1-10) ____
SCORE ____

Hillrock Estate

Proof: 90.6
Age: NAS
Type: Not Straight
Style: Special Finish
Mash Bill: Not available
Color: Straw
Price: $$$$

Mike

Nose: Honey, vanilla, graham crackers.
Taste: Berries with caramel and oak. A bit thin sweet in the mouth.
Finish: Short and thin – caramel and oak.

Susan

Nose: Caramel apple with some pear, baking spice, touch of anise.
Taste: Caramel candy bar with some sweet spice.
Finish: Long and peppery. Water tames the pepper.

Notes

Aged using the Solara method. From Hillrock Estate Distillery, Hudson Halley, New York. According to the label, the distillery floor malts its grains.

My Score

Notes: ____________________________________

____

____

Appearance (1-5) ____ Nose (1-25) ____
Taste (1-25) ____ Finish (1-25) ____
Complexity (1-10) ____ Overall Impression (1-10) ____
SCORE ____

Historic B. Bird Bourbon

Proof:	100
Age:	At least 2 years
Type:	Straight
Style:	Traditional
Mash Bill:	Not available
Color:	Light Straw
Price:	$$$

Mike

Nose: Very light nose. Vanilla with a little corn and oak.

Taste: Vanilla and corn with cardamom, nutmeg, and a little note of honey and oak.

Finish: Medium long with oak and spice.

Susan

Nose: Very light with suggestions of vanilla, corn, and mint.

Taste: Brown sugar, a dusting of cardamom. Pleasant, but not complex.

Finish: Fairly long with a peppery ending.

Notes

Sourced from MGP and bottled by Kentucky Artisan Distillers. Quite nice for its age.

My Score

Notes: ______________________________

____

____

Appearance (1-5) ____ Nose (1-25) ____

Taste (1-25) ____ Finish (1-25) ____

Complexity (1-10) ____ Overall Impression (1-10) ____

SCORE ____

Hooker's House

Proof: 100
Age: 6 years
Type: Straight
Style: Special Finish
Mash Bill: High rye
Color: Light Amber
Price: $$$

Mike

Nose: Caramel and fruit with some tobacco and leather.
Taste: Caramel and berries with nutmeg and cinnamon.
Finish: Long and spicy with some oak and fruit.

Susan

Nose: Vanilla, caramel, orchard fruit.
Taste: Caramel and ripe berries. Touch of baking spices.
Finish: Peppery. Water releases fruit.

Notes

Finished in pinot noir barrels and bottled by Prohibition Spirits, Sonoma, California. Distilled in Indiana. (Probably MGP, but not stated.)

My Score

Notes: ____________________________________

____

____

Appearance (1-5) ____ Nose (1-25) ____
Taste (1-25) ____ Finish (1-25) ____
Complexity (1-10) ____ Overall Impression (1-10) ____

SCORE ____

Hotel Tango Reserve Bourbon

Proof: 90
Age: 6 years
Type: Straight
Style: Traditional
Mash Bill: 74% corn, 22% rye, 4% malted barley
Color: Dark Straw
Price: $$$

Mike

Nose: Corn, vanilla, and berries with a hint of baking spice and oak.
Taste: Corn, blackberries, raspberries, vanilla, oak, and a little allspice.
Finish: Sweet with lingering fruit, oak, and baking spices.

Susan

Nose: Cinnamon rolls with vanilla icing and a little baked apple.
Taste: Crème brulee sprinkled with allspice and brown sugar and a squeeze of berry juice. Rich mouthfeel.
Finish: Sweet and warm.

Notes

Sourced from MGP and bottled by Kentucky Artisan Distillers. Limited series 002. There are more. Quite nice for its age.

My Score

Notes: ______________________________
____
____

Appearance (1-5) ____ Nose (1-25) ____
Taste (1-25) ____ Finish (1-25) ____
Complexity (1-10) ____ Overall Impression (1-10) ____
SCORE ____

Hudson Baby Bourbon

Proof: 90.7
Age: Under 4 years
Type: Not Straight
Style: Small Batch, Craft
Mash Bill: Not available
Color: Dark Straw
Price: $$$

Mike

Nose: Very light. Oak and corn with some yeast dough. Very light floral note.
Taste: Oak tannins and corn. Hint of pepper spice. Not complex.
Finish: Long and dry with oak tannins and spice. Water tames the dryness, but makes it less interesting.

Susan

Nose: Corn and oak. A little vanilla as it sits in the glass.
Taste: Cinnamon, some corn. Spice dominates. Not complex.
Finish: Long and hot. Ouch.

Notes

Batch 11 of 2013, Bottle 3230. Tuthilltown Spirits, Gardiner, New York. Sold in 375ml bottles. Uses corn grown in New York state.

My Score

Notes: ____________________________________

____

____

Appearance (1-5) ____ Nose (1-25) ____
Taste (1-25) ____ Finish (1-25) ____
Complexity (1-10) ____ Overall Impression (1-10) ____
SCORE ____

Iowa Distilling Straight Bourbon Whiskey

Proof: 90
Age: 2 years
Type: Straight
Style: Traditional
Mash Bill: Not available
Color: Straw
Price: $$

Mike

Nose: Apricots and dates up front with vanilla and oak and a hint of fine leather.
Taste: Vanilla and a bit of anise or black licorice with some apricots and cherries and oak.
Finish: Long and fruity. Very nice ending with well-balanced dry oak.

Susan

Nose: Corn flakes with some caramel and oak.
Taste: Vanilla, sweet mint, and white pepper.
Finish: Peppery with vanilla at the very end.

Notes

Batch No. 7, Bottle No. 224.

My Score

Notes: ____

____

____

Appearance (1-5) ____ Nose (1-25) ____
Taste (1-25) ____ Finish (1-25) ____
Complexity (1-10) ____ Overall Impression (1-10) ____

SCORE ____

Ironroot Harbinger

Proof:	115
Age:	27 months
Type:	Straight
Style:	Pot Distilled
Mash Bill:	Purple, Bloody Butcher, flint, and yellow dent corn, plus rye
Color:	Light Amber
Price:	$$$

Mike

Nose: French vanilla and berries.

Taste: Corn, vanilla, raspberries, cherries with a little pepper since.

Finish: Long and dry with oak and pepper.

Susan

Nose: Buttered roasted corn with vanilla, almonds, and peaches.

Taste: Corn and oak with a good dash of pepper. Water brings out vanilla and fruit.

Finish: Long and both sweet and spicy.

Notes

Flagship expression. Also look for the 103 proof Promethean.

My Score

Notes: ______________________________

____

____

Appearance (1-5) ____ Nose (1-25) ____

Taste (1-25) ____ Finish (1-25) ____

Complexity (1-10) ____ Overall Impression (1-10) ____

SCORE ____

J. Henry & Sons Bellefontaine Reserve

Proof:	105
Age:	5 years
Type:	Straight Bourbon
Style:	Special Finish, Four Grain
Mash Bill:	60% corn, 14% wheat, 14% rye, 12% malted barley
Color:	Light Amber
Price:	$$$

Mike

Nose: Vanilla and cotton candy, raisins and a hint of oak.

Taste: Fruit – raisins and berries – with caramel and spice.

Finish: Very peppery and dry with oak tannins.

Susan

Nose: Honeyed fruit with sweet mint and caramel corn.

Taste: Rich mouthfeel with ripe apples, marzipan, and sweet spices.

Finish: Pepper drying to sweet oak.

Notes

Family farm distillery in Dane, Wisconsin. Red heirloom corn, rye, and wheat grown on the farm are used in the bourbon. Finished for 4-6 months in cognac casks. Available in Wisconsin, Minnesota, and Chicago area. Bottle tasted was no. 78 of batch no. 2. Water softens and sweetens without diminishing.

My Score

Notes: ______________________________

____

____

Appearance (1-5)	____	Nose (1-25)	____
Taste (1-25)	____	Finish (1-25)	____
Complexity (1-10)	____	Overall Impression (1-10)	____

SCORE ____

J. Henry & Sons Cask Strength

Proof: 125.2
Age: 5 years
Type: Straight Bourbon
Style: Barrel Strength, Four Grain, Small Batch
Mash Bill: 60% corn, 14% wheat, 14% rye, 12% malted barley
Color: Straw
Price: $$$

Mike

Nose: Vanilla and corn with some leather and oak.
Taste: Citrus, vanilla, and pepper.
Finish: Long, citrus notes and spice. Water makes it very peppery.

Susan

Nose: Fruit, hard candy, crème brulee.
Taste: Sweet on the tongue. Bananas and vanilla.
Finish: Long and fruity with a lingering warmth.

Notes

Water brings out vanilla on the nose and pepper on the palate. Strangely, does not amplify fruit as is often the case with cask strength bourbons. Small Batch No. 2, Bottle 629

My Score

Notes: ____

____

____

Appearance (1-5) ____ Nose (1-25) ____
Taste (1-25) ____ Finish (1-25) ____
Complexity (1-10) ____ Overall Impression (1-10) ____

SCORE ____

J. Henry & Sons Straight Bourbon

Proof: 92
Age: 5 years
Type: Straight Bourbon
Style: Small Batch, Four Grain
Mash Bill: 60% corn, 14% wheat, 14% rye, 12% malted barley
Color: Straw
Price: $$$

Mike

Nose: Vanilla and cotton candy with a hint of oak.
Taste: Vanilla and corn, some baking spices and oak.
Finish: Starts sweet but dries out with oak.

Susan

Nose: Sweet mint, roasted corn, sweet spices.
Taste: Fresh cereal, anise, bit of apple.
Finish: Both sweet and peppery. Nicely balanced.

Notes

From Henry Farms Prairie Spirits in Dane, Wisconsin. Bourbons from this distillery are available in Wisconsin, Minnesota, and Chicago area. Corn is an heirloom red.

My Score

Notes: ______________________________

____

____

Appearance (1-5) ____ Nose (1-25) ____
Taste (1-25) ____ Finish (1-25) ____
Complexity (1-10) ____ Overall Impression (1-10) ____
SCORE ____

J. Riddle Peated Bourbon

Proof: 91
Age: NAS
Type: Straight
Style: Peated
Mash Bill: 79% Michigan corn, 21% Scottish barley
Color: Pale Straw
Price: $$

Mike

Nose: Peat and corn bread batter with a hint of vanilla.
Taste: Peat, corn, vanilla, a hint of fruit.
Finish: Long and peaty.

Susan

Nose: Peat and corn with lots of saddle leather.
Taste: Barnyard (in a good way) with smoke and leather.
Finish: Spicy!

Notes

Two James Spirits Company, Detroit and Grand Rapids, Michigan. Might this lure some Scotch drinkers to bourbon? Nice change of pace.

My Score

Notes: ______________________________

____

____

Appearance (1-5) ____ Nose (1-25) ____
Taste (1-25) ____ Finish (1-25) ____
Complexity (1-10) ____ Overall Impression (1-10) ____

SCORE ____

James E. Pepper 1776

Proof: 100
Age: NAS
Type: Straight
Style: Small Batch
Mash Bill: 38% rye
Color: Dark Straw
Price: $$

Mike

Nose: Corn and vanilla with some oak. Old leather and nutmeg.
Taste: Candied fruit – berries and cherries some pepper spice, leather or tobacco and oak.
Finish: Sweet to start, but dries out nicely with some oak tannins.

Susan

Nose: Caramel and a little licorice. Some faint mint.
Taste: Vanilla, licorice, sweet spice. Adding with water releases nuttiness, but no fruit.
Finish: Starts sweet and ends with spice Not as long as expected from the proof.

Notes

Jas. E. Pepper and Co., Bardstown, Kentucky. But sourced from elsewhere, possibly MGP in Lawrenceberg, Indiana.

My Score

Notes: ______________________________

____

____

Appearance (1-5) ____ Nose (1-25) ____
Taste (1-25) ____ Finish (1-25) ____
Complexity (1-10) ____ Overall Impression (1-10) ____

SCORE ____

Jeffer's Creek

Proof: 80
Age: 6 years
Type: Straight Bourbon
Style: Small Batch
Mash Bill: Proprietary
Color: Pale Light Straw
Price: $

Mike

Nose: Not complex. Vanilla and a hint of leather and spice – anise?
Taste: Not hot at all. Sweet vanilla, liquorice and some other sweet spice – cinnamon – and some oak.
Finish: Medium long and turns dry with oak and some spice.

Susan

Nose: Candy corn, light wood smoke, not complex.
Taste: Wood smoke, caramel, honey.
Finish: Short, but warm with a little sweet spice at the end.

Notes

Label states that it is bottled by King's Mark Co. of Louisville. Might be more interesting at higher proof.

My Score

Notes: ______________________________

____

____

Appearance (1-5) ____ Nose (1-25) ____
Taste (1-25) ____ Finish (1-25) ____
Complexity (1-10) ____ Overall Impression (1-10) ____
SCORE ____

Jefferson's

Proof: 82.3
Age: NAS (website states 8-12 years)
Type: Straight
Style: Small Batch
Mash Bill: Not available
Color: Light Straw
Price: $$

Mike

Nose: Corn and vanilla with a hint of tobacco and oak.
Taste: Corn and vanilla with some sweet spice. Clove? A bit of apricot and oak.
Finish: Starts sweet with the caramel/spice, but dries out some with oak tannins. Not overly long.

Susan

Nose: Sweet corn, vanilla, some oak.
Taste: Corn, vanilla and sweet oak. Not complex, but has a rich, creamy mouthfeel.
Finish: Medium length. Dry, and smooth

Notes

Uncomplicated bourbon makes a pleasant, simple sip. Serve neat or with a single ice cube. Sourced from unknown distillery. Distributed by Castle Brands.

My Score

Notes: ______________________________

____

____

Appearance (1-5) ____ Nose (1-25) ____
Taste (1-25) ____ Finish (1-25) ____
Complexity (1-10) ____ Overall Impression (1-10) ____

SCORE ____

Jefferson's Ocean

Proof: 90
Age: NAS
Type: Straight
Style: Small Batch
Mash Bill: Not available
Color: Light Straw
Price: $$$

Mike

Nose: Vanilla, oak, and sea salt. Very light nose.
Taste: Vanilla and corn with some ocean air – salt and iodine. Some oak spice.
Finish: Long and dry with oak and sea salt.

Susan

Nose: Hard candy, vanilla, ripe cherries and oak.
Taste: Light vanilla and, believe it or not, a little sea salt. Sweet caramel and a touch of orchard fruit. Very smooth.
Finish: Long and a little salty. Unexpectedly pleasant change from pepper.

Notes

Put on a ship and aged at sea. This was from Voyage Three, Barrel 54. A gimmick? Perhaps, but the light salt aromas are certainly present. Very interesting.

My Score

Notes: ______________________________

____

____

Appearance (1-5) ____ Nose (1-25) ____
Taste (1-25) ____ Finish (1-25) ____
Complexity (1-10) ____ Overall Impression (1-10) ____

SCORE ____

Jefferson's Presidential Select

Proof: 94
Age: 18 years
Type: Straight
Style: Small Batch, Wheated
Mash Bill: Uses wheat in place of rye
Color: Light Amber Red
Price: $$$

Mike

Nose: Caramel and apples with some hazelnuts. Not complex, but full-bodied.

Taste: Caramel, chocolate and a bit of sweet spice – cinnamon and nutmeg. Fine tobacco and honey.

Finish: Long, sweet and pleasant. Not complex, but flavorful.

Susan

Nose: Apricot and dried mint. A little new leather.

Taste: Dried fruit, toffee, caramel with hazelnuts and buttered popcorn.

Finish: Long and smooth. Initial fruit gives way to dry oak at the very end.

Notes

Very limited release of bottling from the closed Stitzel-Weller Distillery in Louisville. We tasted the 18-year-old. The 17-year-old is pictured. Distributed by Castle Brands

My Score

Notes: ______________________________

____

____

Appearance (1-5) ____ Nose (1-25) ____

Taste (1-25) ____ Finish (1-25) ____

Complexity (1-10) ____ Overall Impression (1-10) ____

SCORE ____

Jefferson's Reserve

Proof: 90.2
Age: 15 years
Type: Straight
Style: Small Batch
Mash Bill: Not available
Color: Light Straw
Price: $$

Mike

Nose: Sweet caramel corn with fine tobacco and leather.
Taste: Caramel and fruit – apricot. A bit of pepper spice and tobacco.
Finish: Starts sweet but dries out. Fairly short. I hoped for a better finish.

Susan

Nose: Caramel, vanilla, toffee and note of new leather.
Taste: Candied corn, caramel and sweet spices with whiff of oak.
Finish: Starts sweet and ends dry and smooth. No heat.

Notes

Works very well in a classic Manhattan. Distributed by Castle Brands

My Score

Notes: ______________________________

____

____

Appearance (1-5) ____ Nose (1-25) ____
Taste (1-25) ____ Finish (1-25) ____
Complexity (1-10) ____ Overall Impression (1-10) ____

SCORE ____

Jefferson's Reserve Groth Reserve Cask Finish

Proof: 90.2
Age: NAS
Type: Straight
Style: Special Finish, Small Batch
Mash Bill: Not available
Color: Light Straw
Price: $$$

Mike

Nose: Fruity. Berries and vanilla. Grape popsicle.
Taste: Grapes and vanilla with sweet spice and oak.
Finish: Starts sweet and get some nice oak and spice.

Susan

Nose: Sweet wine notes. Fruit dominates with some sweet vanilla.
Taste: Wine, vanilla, cinnamon.
Finish: Long and warm and fruity.

Notes

Very small batch. Bottle 3711 of batch No. 1. Finished in French oak Groth cabernet sauvignon barrels. Might be a very good sip with pork tenderloin.

My Score

Notes: ______________________________

____

____

Appearance (1-5)	____	Nose (1-25)	____
Taste (1-25)	____	Finish (1-25)	____
Complexity (1-10)	____	Overall Impression (1-10)	____

SCORE ____

Jefferson's Reserve Pritchard Hill Cabernet Cask Finish

Proof: 90.2
Age: NAS
Type: Finished
Style: Straight
Mash Bill: Not available
Color: Light Amber
Price: $$$

Mike

Nose: Vanilla, raisins, and berries with a hint of chocolate and oak.
Taste: Vanilla, raspberry, and blackberry with some baking spices and oak.
Finish: Long and dry with some fruit and lots of oak.

Susan

Nose: Sweet dates, caramel, crème brulee.
Taste: Caramel, brown sugar, dates, figs, ripe apples with some baking spices.
Finish: Long and sweet and the wine note really asserts itself here.

Notes

One of several Jefferson's limited releases. Kentucky Artisan Distillery, Crestwood, Kentucky, https://jeffersonsbourbon.com/jeffersons-reserve/.

My Score

Notes: ______________________________
____
____

Appearance (1-5) ____ Nose (1-25) ____
Taste (1-25) ____ Finish (1-25) ____
Complexity (1-10) ____ Overall Impression (1-10) ____

SCORE ____

Jeptha Creed Bottled-in-Bond

Proof: 100
Age: At least 4 years
Type: Straight
Style: Bonded, Farm Distilled
Mash Bill: 75% Bloody Butcher corn, 20% malted rye, 5% malted barley
Color: Light Amber
Price: $$

Mike

Nose: Nice aroma. Caramel and pears with a little spice and oak. Opens as it breathes. To more spice, anise, and cardamom.
Taste: Caramel, spice, pepper, cardamom, oak, and a little berry fruit.
Finish: Long with oak and pepper.

Susan

Nose: Old books with some potpourri and light oak. Caramel dance around the edges.
Taste: Caramel corn and herbal spices with some chopped nuts.
Finish: Lots of pepper zips about on the tongue. Long with a sweet ending.

Notes

Give it some time to open in the glass for maximum enjoyment.

My Score

Notes: ______________________________

____

____

Appearance (1-5) ____ Nose (1-25) ____
Taste (1-25) ____ Finish (1-25) ____
Complexity (1-10) ____ Overall Impression (1-10) ____
SCORE ____

Jeptha Creed Four Grain Bourbon

Proof: 98
Age: 2 years
Type: Straight
Style: Four Grain, Farm Distilled
Mash Bill: 70% Bloody Butcher corn, 15% malted rye, 10% malted wheat, 5% malted barley
Color: Dark Straw
Price: $$

Mike

Nose: Corn, vanilla, and malted aroma with a little peach or apricot and oak.
Taste: Corn with some lemon zest, parches, and cinnamon with a hint of oak.
Finish: Medium long with citrus and oak.

Susan

Nose: Roasted corn, some peaches, cinnamon, and vanilla with some maltiness.
Taste: Vanilla and lemons, candy corn, nutmeg, and sweet cherries.
Finish: Warm with pepper oak, and lingering citrus.

Notes

Jeptha Creed Distillery, Shelbyville, Kentucky. A drop or two of water amplifies the citrus and baking spices.

My Score

Notes: ________________________________

____

____

Appearance (1-5) ____ Nose (1-25) ____
Taste (1-25) ____ Finish (1-25) ____
Complexity (1-10) ____ Overall Impression (1-10) ____
SCORE ____

Jesse James America's Outlaw Bourbon

Proof: 80
Age: At least 36 months
Type: Not Straight
Style: Traditional
Mash Bill: Not available
Color: Pale Straw
Price: $

Mike

Nose: Very light and young. Not much here but corn and vanilla.
Taste: Very light. Sweet corn vanilla. Hint of oak.
Finish: SShort and slightly dry with oak Tannins.

Susan

Nose: Faint vanilla and a little cinnamon. Very light.
Taste: Smooth and sweet on the tongue. Light caramel, a little honey. No bite.
Finish: Very short, sweet.

Notes

Londonderry, New Hampshire, but distilled in Kentucky by an unnamed distiller. Singer Jesse James Dupree founded the brand. Pretty mild to be named after a tough guy.

My Score

Notes: ______________________________
____
____

Appearance (1-5) ____ Nose (1-25) ____
Taste (1-25) ____ Finish (1-25) ____
Complexity (1-10) ____ Overall Impression (1-10) ____
SCORE ____

Jethro T. Boots

Proof: 80
Age: NAS
Type: Straight
Style: Traditional
Mash Bill: Not available
Color: Very Light Straw
Price: $

Mike

Nose: Candy corn with a hint of oak.
Taste: Sweet honey and pepper. Hint of fruit.
Finish: Starts sweet and dries to light oak tannins.

Susan

Nose: Simple. Light vanilla, sweet corn.
Taste: Sweet corn, vanilla, faint candied apple.
Finish: Surprisingly long and warm. Oak at the end.

Notes

Bottled by the Legacy Distilling Company, Louisville, Kentucky. Sourced from Barton 1792 in Bardstown, Kentucky.

My Score

Notes: ______________________________

____

____

Appearance (1-5) ____ Nose (1-25) ____
Taste (1-25) ____ Finish (1-25) ____
Complexity (1-10) ____ Overall Impression (1-10) ____
SCORE ____

Joe Louis

Proof: 90
Age: 2 years
Type: Straight
Style: Pot Distilled
Mash Bill: 66% corn, 14% rye, 20% malted barley
Color: Straw
Price: $$

Mike

Nose: Corn and vanilla with a little spice and oak.
Taste: Corn and vanilla with a little orange zest, cinnamon, and oak.
Finish: Medium long with oak, cinnamon, and lingering corn.

Susan

Nose: Raw corn, faint spice, a little apple peel.
Taste: Grain forward, but caramel and spice emerge quickly.
Finish: Nice note punctuates the end.

Notes

Bottled by Davis Valley Distillery, Rural Retreat, Virginia.

My Score

Notes: ____

____

____

Appearance (1-5) ____ Nose (1-25) ____
Taste (1-25) ____ Finish (1-25) ____
Complexity (1-10) ____ Overall Impression (1-10) ____
SCORE ____

John B. Stetson

Proof:	84
Age:	4 years
Type:	Straight
Style:	Pot Stilled
Mash Bill:	Not available
Color:	Light Straw
Price:	$$

Mike

Nose: Corn and yeast. Very light.
Taste: Candy apple with some pepper spice, a hint of oak and tobacco. A bit hot.
Finish: Long and peppery.

Susan

Nose: Very light. Vanilla and corn.
Taste: Vanilla, some apple and honey. Black pepper.
Finish: Fades quickly to black pepper.

Notes

Frank-Lin Distillers, Fairfield, California. This facility sources bourbon from undisclosed distilleries in Kentucky and bottles it under several labels.

My Score

Notes: ______________________________

____

____

Appearance (1-5) ____ Nose (1-25) ____
Taste (1-25) ____ Finish (1-25) ____
Complexity (1-10) ____ Overall Impression (1-10) ____

SCORE ____

J.R. Ewing

Proof:	80
Age:	4 years
Type:	Straight
Style:	Traditional
Mash Bill:	Not available
Color:	Light Straw
Price:	$$$

Mike

Nose: Very light. Very yeasty – bread dough or malt. Hint of vanilla and corn.

Taste: Very light. Slightly sweet with some caramel and light fruit – pears and apricots.

Finish: Short and sweet. Not much to it. Corn sweetness on the finish.

Susan

Nose: Very light, biscuit. With touch of corn. Not complex.

Taste: Light and sweet. Vanilla and pear.

Finish: Sweet. Short.

Notes

Made by Strong Spirits of Bardstown, Kentucky for Southfork Bottling Company, Dallas, Texas.

My Score

Notes: ____________________________________

____

____

Appearance (1-5) ____ Nose (1-25) ____

Taste (1-25) ____ Finish (1-25) ____

Complexity (1-10) ____ Overall Impression (1-10) ____

SCORE ____

Kentucky Owl

Proof: 118.4
Age: NAS
Type: Straight
Style: Small Batch, Barrel Proof
Mash Bill: Not available
Color: Dark Amber
Price: $$$$

Mike

Nose: Caramel and molasses with some nutmeg, Fine leather and tobacco. Water brings out some berries and dates.
Taste: Candies cherries and caramel with some sweet spice, tobacco, and oak. Water brings out caramel, but lessens fruit.
Finish: Long and dry with oak tannins. Water makes the finish more tannic.

Susan

Nose: Vanilla, caramel, sweet oak, cinnamon, cherries. Very rich. I could nose this all day.
Taste: Rich and smooth mouthfeel. Water reveals cinnamon, cherries and lots of vanilla.
Finish: Long and rich, with cinnamon, vanilla and lingering caramel.

Notes

A mingling of cask strength bourbons sourced from more than one Kentucky distillery. This is a resurrection of a brand the Dedman family (owners of Beaumont Inn in Harrodsburg, Kentucky) made at their pre-prohibition distillery. Batch No. 1. Bottle 1235/1506.

My Score

Notes: ____________________

____

____

Appearance (1-5) ____ Nose (1-25) ____
Taste (1-25) ____ Finish (1-25) ____
Complexity (1-10) ____ Overall Impression (1-10) ____

SCORE ____

Kentucky Owl Confiscated

Proof: 96.4
Age: NAS
Type: Straight
Style: Traditional
Mash Bill: Not available
Color: Straw
Price: $$$$

Mike

Nose: Toasted marshmallows, oak, and some ripe apple and banana.
Taste: Oak forward with vanilla, pepper spice, and banana pudding with a hint of green apple.
Finish: Long and spicy.

Susan

Nose: Caramel, cherries, peaches, and cinnamon.
Taste: Caramel softens to vanilla with more cherry, some banana, and lots of baking spices.
Finish: Warm and spicy.

Notes

Kentucky Owl, LLC, Bardstown. Now owned by the Stoli Group. Sourced from undisclosed distillery.

My Score

Notes: ____________________

____

____

Appearance (1-5) ____ Nose (1-25) ____
Taste (1-25) ____ Finish (1-25) ____
Complexity (1-10) ____ Overall Impression (1-10) ____

SCORE ____

Kentucky Owl Takumi Edition

Proof: 100

Age: Blend of 4, 5, 6, and 13 year bourbons

Type: Straight

Style: Four Grain, Limited Edition

Mash Bill: Undisclosed proportions of bourbons containing corn, rye, wheat, malted barley

Color: Straw

Price: $$$$

Mike

Nose: Rich caramel and chocolate with oak.

Taste: Caramel, ripe apples, allspice, oak.

Finish: Medium long with oak and spice.

Susan

Nose: Milk chocolate, apple peel, crème brulee, light baking spice.

Taste: Dark fruit, leather, tobacco, oak.

Finish: Long, drying to oak and pepper.

Notes

Collaboration between Kentucky Owl master blender John Rhea and Nagahoma Distillery master blender Yahisa Yusuke; 25,000 bottles released.

My Score

Notes: ______________________________

____

____

Appearance (1-5) ____ Nose (1-25) ____

Taste (1-25) ____ Finish (1-25) ____

Complexity (1-10) ____ Overall Impression (1-10) ____

SCORE ____

Kings County

Proof: 90
Age: 1-2 years
Type: Not Straight
Style: Peated, Craft
Mash Bill: 70% corn, 30% peated malt
Color: Light Straw
Price: $$$

Mike

Nose: Peat, vanilla, and corn. Dominated by peat.
Taste: Peak, sweet corn, and vanilla. Peat dominates.
Finish: Long and peaty. Not complicated. Too much like Scotch. Not enough like bourbon.

Susan

Nose: Grass clippings, oak, a whiff of corn.
Taste: Smoke and more smoke. Some sweet.
Finish: Long and smoky. One-dimensional.

Notes

Made by Kings County Distillery, Brooklyn, New York. Located in the Brooklyn Navy Yard. Sold in 200ml bottles for $25 or so. Price works out to over $90/fifth.

My Score

Notes: ______________________________
____
____

Appearance (1-5) ___ Nose (1-25) ___
Taste (1-25) ___ Finish (1-25) ___
Complexity (1-10) ___ Overall Impression (1-10) ___
SCORE ___

Koval

Proof: 94
Age: Less than 4 years
Type: Not Straight
Style: Single Barrel
Mash Bill: Not available
Color: Light Straw
Price: $$

Mike

Nose: Chemical – wormwood and herbs.
Taste: Sweet honey and fruit. Raisins and berries.
Finish: Very dry. Almost astringent.

Susan

Nose: Herbal and slightly resinous.
Taste: Pepper dominates vanilla and orchard fruit.
Finish: Long and peppery with a little mint at the end.

Notes

The taste is better than the nose. This could make a very interesting Manhattan. From Koval Distillery, Chicago, Illinois.

My Score

Notes: ______________________________________

____

____

Appearance (1-5) ____ Nose (1-25) ____
Taste (1-25) ____ Finish (1-25) ____
Complexity (1-10) ____ Overall Impression (1-10) ____

SCORE ____

Landry Stakes

Proof: 90
Age: 4 years
Type: Straight
Style: Small Batch, Wheated
Mash Bill: 70% corn, 20% wheat, 10% malted barley
Color: Straw
Price: $$$$

Mike

Nose: Light nose. Vanilla and corn with some apricot and oak.
Taste: Vanilla, corn, a little orange peel, allspice, and oak.
Finish: Medium long with oak and allspice.

Susan

Nose: Lightly roasted corn with brown sugar, a whiff of cinnamon, and some faint pear.
Taste: Brown sugar dominates with some peppery tannins, a little ginger and orchard fruit.
Finish: Long and dry.

Notes

Balanced, but not complex. Produced by Crittenden Distillery, Kiln, Mississippi.

My Score

Notes: ______________________________________

____

____

Appearance (1-5) ____ Nose (1-25) ____
Taste (1-25) ____ Finish (1-25) ____
Complexity (1-10) ____ Overall Impression (1-10) ____

SCORE ____

Leadslingers

Proof: 80
Age: NAS
Type: Not Straight
Style: Small Batch
Mash Bill: Not available
Color: Very Faint Straw
Price: $$

Mike

Nose: Corn and not much else. Hint of faint vanilla.
Taste: Thin but sweet, with vanilla and corn.
Finish: Short and sweet.

Susan

Nose: Light leather, corn, alcohol.
Taste: A little caramel and butterscotch.
Finish: Short, oaky.

Notes

From Scissortail Distillery, Moore, Oklahoma.

My Score

Notes: ______________________________________
____
____

Appearance (1-5) ____ Nose (1-25) ____
Taste (1-25) ____ Finish (1-25) ____
Complexity (1-10) ____ Overall Impression (1-10) ____

SCORE ____

Lee W. Sinclair

Proof: 90
Age: 2 years
Type: Straight
Style: Pot Distilled, Four Grain
Mash Bill: 60% corn, 13% wheat, 17% oats, 10% caramel malted barley
Color: Straw
Price: $$

Mike

Nose: Corn and vanilla with some apple fruit, baking spice, and oak.
Taste: Corn, vanilla, apples, spices – pepper and a hint of cinnamon – oak.
Finish: Long and spicy with a hint of oak.

Susan

Nose: Corn and multigrain cereal with a little bit of vanilla.
Taste: Buttered popcorn, vanilla wafers, some ripe apples. Nutty.
Finish: Spicy finish drying to oak.

Notes

Distilled by Spirits of French Lick, West Baden, Indiana. One of several expression made by the distillery.

My Score

Notes: ______________________________
____
____

Appearance (1-5) ___ Nose (1-25) ___
Taste (1-25) ___ Finish (1-25) ___
Complexity (1-10) ___ Overall Impression (1-10) ___
SCORE ___

Leiper's Fork Bottled-in-Bond

Proof: 100
Age: 5 years
Type: Straight
Style: Pot Distilled, Bonded, Wheated
Mash Bill: 70% corn, 15% wheat, 15% malted barley
Color: Dark Straw
Price: $$$

Mike

Nose: Sorghum, prunes and dates, with a little baking spice and oak.
Taste: Caramel and sorghum, plums and prunes with some dates, cinnamon, nutmeg, and oak.
Finish: Oak and spice with a lingering fruity sweetness.

Susan

Nose: Very cognac-like with raisins and dates followed by rich caramel and some milk chocolate.
Taste: Chocolate, and caramel with molasses and dark fruit. Rich mouthfeel.
Finish: Full complex of flavors carry through to the end with raisins and spice.

Notes

Leiper's Fork Distillery, Franklin, Tennessee. Uses toasted barley to enhance flavor. Low barrel entry proof of 110. Makes other expressions of bourbon and Tennessee whiskey.

My Score

Notes: ____________________________________

____

____

Appearance (1-5) ____ Nose (1-25) ____
Taste (1-25) ____ Finish (1-25) ____
Complexity (1-10) ____ Overall Impression (1-10) ____

SCORE ____

Lexington

Proof: 86
Age: NAS
Type: "Straight" not on label
Style: Traditional
Mash Bill: Proprietary
Color: Light Straw
Price: $$

Mike

Nose: Very light. Sweet corn. A little vanilla. A little floral note.
Taste: Thin and watery. Corn and a little vanilla.
Finish: Not much. Very short and sweet. Brown vodka.

Susan

Nose: Light. Caramel, apple, sweet corn.
Taste: Very light. Some sweet spice and peaches with the vanilla.
Finish: Very short. Practically none.

Notes

Named for a famed racehorse, who appears on the label. Bottled by Western Spirits Beverage Co., Bowling Green, Kentucky. Kentucky-sourced.

My Score

Notes: ____________________________________

____

____

Appearance (1-5) ____ Nose (1-25) ____
Taste (1-25) ____ Finish (1-25) ____
Complexity (1-10) ____ Overall Impression (1-10) ____
SCORE ____

Linkumpinch Bourbon Bottled-in-Bond

Proof: 100
Age: 4 years
Type: Straight
Style: Small Batch, Bonded
Mash Bill: Not available
Color: Light Amber
Price: $$$

Mike

Nose: Simple. Vanilla, apricots, oak.
Taste: Vanilla, apricots, cinnamon, nutmeg, oak.
Finish: Long with oak and spice and lingering vanilla sweetness.

Susan

Nose: Multigrain aroma followed by apple peel and light caramel.
Taste: Dark fruit, pepper, oak.
Finish: Touch of chocolate at the very end.

Notes

Dueling Grounds Distillery, Franklin, Kentucky.

My Score

Notes: ______________________________
____
____

Appearance (1-5) ____ Nose (1-25) ____
Taste (1-25) ____ Finish (1-25) ____
Complexity (1-10) ____ Overall Impression (1-10) ____

SCORE ____

Linkumpinch Bourbon Single Barrel Cask Strength

Proof: 122.6
Age: Minimum 6 years
Type: Straight
Style: Single Barrel, Cask Strength
Mash Bill: Not available
Color: Light Amber
Price: $$$

Mike

Nose: Vanilla, apricots, baking spice, oak.
Taste: Vanilla, apricots and berries, cinnamon and nutmeg, oak.
Finish: Very long and dry with oak and spice.

Susan

Nose: Chocolate covered nuts, very light cinnamon, peaches.
Taste: Nutty with caramel, some orange peel, and peppery oak.
Finish: Long and peppery, but not hot.

Notes

Dueling Grounds Distillery, Franklin, Kentucky. Water brings out more caramel and fruit.

My Score

Notes: ______________________________

____

____

Appearance (1-5) ____ Nose (1-25) ____
Taste (1-25) ____ Finish (1-25) ____
Complexity (1-10) ____ Overall Impression (1-10) ____
SCORE ____

Litchfield Distillery Batcher's Cask Strength

Proof: 122
Age: 4 years
Type: Straight
Style: Barrel Proof
Mash Bill: 70% corn, 25% rye, 5% malted barley
Color: Straw
Price: $$$

Mike

Nose: Very light nose. Vanilla and pears, a hint of spice and oak.
Taste: Vanilla and pears with a little hazelnut. Pear tart! With nutmeg, allspice, and oak.
Finish: Medium ling with oak and spice.

Susan

Nose: Subtle. A touch of spice, pear, and vanilla.
Taste: Hard candy with pear and light pepper. And baking spices.
Finish: Short and dry.

Notes

Water brings out the spice, oak, and fruit. Other bourbon expressions have included one at 92 proof, a double-barreled, and one finished in port wine casks. Litchfield Distillery, Litchfield, Connecticut.

My Score

Notes: ______________________________
____
____

Appearance (1-5) ____ Nose (1-25) ____
Taste (1-25) ____ Finish (1-25) ____
Complexity (1-10) ____ Overall Impression (1-10) ____
SCORE ____

Luca Mariano Old Americana

Proof: 83
Age: 4 years
Type: Straight
Style: Small Batch
Mash Bill: Not available
Color: Light Straw
Price: $$

Mike

Nose: Vanilla and a hint of berries or plums with some oak.
Taste: Vanilla and corn with a little plum, black pepper, and oak.
Finish: Medium long with oak and lots of black pepper.

Susan

Nose: Light and fruity with apples and berries.
Taste: Subtle. Some caramel and fruit. There's little spice, but it's very sippable.
Finish: Ends fairly quickly with just a whisper of oak.

Notes

Sourced from an undisclosed Kentucky distillery. Website cites plans for eventually building the company's own distillery near Danville, Kentucky.

My Score

Notes: ____

____

____

Appearance (1-5) ____ Nose (1-25) ____
Taste (1-25) ____ Finish (1-25) ____
Complexity (1-10) ____ Overall Impression (1-10) ____
SCORE ____

Lythgoe Bourbon Four Grain Founder's Collection

Proof: 94
Age: 9 years
Type: Straight
Style: Single Barrel, Four Grain
Mash Bill: 67% corn, 12.5% rye, 12.5% wheat 8% malted barley
Color: Bronze
Price: $$$

Mike

Nose: Vanilla, corn, peaches, apricots, and oak.
Taste: Vanilla, peaches, cardamom, oak.
Finish: Long with oak, cardamom, and lingering peaches.

Susan

Nose: Multigrain cereal with some apricot and vanilla. Very soft.
Taste: Peach peal, wheat bread, light vanilla, pinch of baking spice.
Finish: Warm and balanced.

Notes

Rich and complex. Distilled at Kentucky Artisan Distillery, Crestwood, Kentucky.

My Score

Notes: ______________________________

____

____

Appearance (1-5) ____ Nose (1-25) ____
Taste (1-25) ____ Finish (1-25) ____
Complexity (1-10) ____ Overall Impression (1-10) ____

SCORE ____

Lythgoe Bourbon Four Grain Master Blender's Collection

Proof: 100.8
Age: 8 years
Type: Straight
Style: Single Barrel, Four Grain
Mash Bill: 67% corn, 12.5% rye, 12.5% wheat 8% malted barley
Color: Dark Amber
Price: $$$

Mike

Nose: Vanilla, caramel, peaches, baking spices, and oak.
Taste: Vanilla, chocolate, peaches, cardamom, oak.
Finish: Long with oak, spice and lingering chocolate.

Susan

Nose: Dark toffee, old leather, dark cherries, pinches of cinnamon and cardamom.
Taste: Toffee, dark chocolate, dark fruit, coffee, baking spice.
Finish: Long, savory, balanced.

Notes

Rich mouthfeel. Distilled at Kentucky Artisan Distillery, Crestwood, Kentucky.

My Score

Notes: ______________________________________

____

____

Appearance (1-5) ____ Nose (1-25) ____
Taste (1-25) ____ Finish (1-25) ____
Complexity (1-10) ____ Overall Impression (1-10) ____

SCORE ____

Mad River Bourbon Whiskey

Proof: 92
Age: Less than 4 years
Type: Not straight
Style: Four Grain
Mash Bill: Corn, wheat, oats, malted barley. Proportions unknown.
Color: Pale Straw
Price: $$

Mike

Nose: Vanilla and peaches with a hint of oak.
Taste: Vanilla and pepper with a hint of peaches.
Finish: Medium long with lots of pepper and vanilla.

Susan

Nose: Buttered popcorn, citrus, vanilla.
Taste: Berries and vanilla with chocolate dancing around the edges of the tongue.
Finish: Smooth and sweet.

Notes

Mad River Distillers, Warren, Vermont.

My Score

Notes: ______________________________________

____

____

Appearance (1-5) ____ Nose (1-25) ____
Taste (1-25) ____ Finish (1-25) ____
Complexity (1-10) ____ Overall Impression (1-10) ____
SCORE ____

Mary Dowling Tequila Finished

Proof: 93
Age: NAS
Type: Straight
Style: Finished
Mash Bill: 70% corn, 25% malted rye, 5% malted barley
Color: Pale Straw
Price: $$$

Mike

Nose: Simple. Vanilla, hint of oak.
Taste: Vanilla, tequila, black pepper, oak.
Finish: Medium long with oak and black pepper.

Susan

Nose: Some berries and light brown sugar.
Taste: Brown sugar, berries, some tequila astringency. Nice mouthfeel.
Finish: Long and warm with a pop of fruit at the end.

Notes

Rabbit Hole Distillery, Louisville, Kentucky. Mary Dowling moved her entire distillery from Kentucky to Mexico during Prohibition, hence the tequila finish as a nod to that history.

My Score

Notes: ______________________________

____

____

Appearance (1-5) ____ Nose (1-25) ____
Taste (1-25) ____ Finish (1-25) ____
Complexity (1-10) ____ Overall Impression (1-10) ____

SCORE ____

MB Roland Straight Bourbon

Proof: 107
Age: More than 2 years
Type: Straight
Style: Small Batch, Non-Chill Filtered
Mash Bill: 78% white corn, 17% rye, 5% malted barley
Color: Straw
Price: $$

Mike

Nose: Vanilla, corn, a hint of crème brulee, and oak.
Taste: Vanilla and corn with pepper spice and a hint of ripe apples.
Finish: Long and dry with oak and a hint of pepper.

Susan

Nose: Roasted corn, vanilla, apples, nuts.
Taste: Lots of vanilla and caramel apple with nutmeg, allspice, a touch of new leather.
Finish: Round and sweet.

Notes

MB Roland Distillery, Pembroke, Kentucky. Farm distillery with a large portfolio of spirits.

My Score

Notes: ____________________________________
____
____

Appearance (1-5) ____ Nose (1-25) ____
Taste (1-25) ____ Finish (1-25) ____
Complexity (1-10) ____ Overall Impression (1-10) ____
SCORE ____

McKenzie

Proof: 91
Age: NAS (website says it is 2 years old)
Type: Not Straight
Style: Craft, Non-Chill Filtered
Mash Bill: Not available
Color: Dark Straw
Price: $$$

Mike

Nose: Very yeasty – bread dough – with corn and vanilla. Some floral notes (roses?) and sweet spices.
Taste: Sweet corn and caramel with some spices – nutmeg and allspice – leather tobacco, pecans.
Finish: Starts sweet, but gets dry with oak tannins. Lasts a long time.

Susan

Nose: Dried corn, ethanol, tobacco opening into yeasty bread dough.
Taste: Sweet corn and caramel. Some sweet spices – cinnamon, allspice.
Finish: Medium long, sweet and smooth a little pepper on the tongue.

Notes

Made at the Finger Lakes Distilling of Burdett, New York. Whiskey is made from locally grown grains.

My Score

Notes: ____

____

____

Appearance (1-5) ____ Nose (1-25) ____
Taste (1-25) ____ Finish (1-25) ____
Complexity (1-10) ____ Overall Impression (1-10) ____

SCORE ____

McKenzie Bottled-in-Bond

Proof: 100
Age: Minimum 4 years
Type: Straight
Style: Bonded, Wheated
Mash Bill: Not available, but contains wheat
Color: Light Straw
Price: $$$$

Mike

Nose: Vanilla and corn with a little apple and honey.
Taste: Vanilla, nutmeg, and cinnamon with apples and oak.
Finish: Nice dry finish with oak and a hint of apple.

Susan

Nose: Corn, alcohol, some brown sugar.
Taste: Corn, some apple, lots of black pepper.
Finish: Starts spicy and on a surprising vanilla note.

Notes

Finger Lakes Distilling, Burdett, New York.

My Score

Notes: ______________________________

____

____

Appearance (1-5) ____ Nose (1-25) ____
Taste (1-25) ____ Finish (1-25) ____
Complexity (1-10) ____ Overall Impression (1-10) ____

SCORE ____

Medley Bros.

Proof: 102
Age: NAS
Type: Straight
Style: Small Batch
Mash Bill: 77% Corn, 10% Rye, 13% Malted Barley
Color: Straw
Price: $$

Mike

Nose: Vanilla and caramel with apricot and fine leather.
Taste: Vanilla and corn with oak wood and fruit – apricots – and tobacco.
Finish: Long and dry with oak tannins and spice.

Susan

Nose: Floral notes with caramel and some nutmeg.
Taste: Sweet corn, a smidge of mint, vanilla.
Finish: Long and warm. Peppery, as one of pepper might expect from the proof.

Notes

Kentucky sourced bourbon bottled by Frank-Lin Distillers, Fairfield, California. A big-shouldered sip. Well-balanced for the proof.

My Score

Notes: ____

____

____

Appearance (1-5) ____ Nose (1-25) ____
Taste (1-25) ____ Finish (1-25) ____
Complexity (1-10) ____ Overall Impression (1-10) ____
SCORE ____

Metze's Select

Proof: 93
Age: 7-9 years
Type: Straight
Style: Small Batch
Mash Bill: See notes
Color: Straw
Price: $$$

Mike

Nose: Caramel and sweet spices with hint of raisins and anise.
Taste: Sweet caramel, berries and dates, some baking spice.
Finish: Starts sweet with fruit. Dries to spicy oak tannins.

Susan

Nose: Caramel, wood smoke, dried apples.
Taste: Baked toffee, sweet mint, baked pears.
Finish: Starts sweet, but jarring bitter note at the end.

Notes

Distilled and released by MGP, Lawrenceburg, Indiana, which otherwise contract distills liquor for independent bottlers. Mingling of 2006 bourbon (21% rye), 3% 2006 bourbon (36% rye), and 59% 2008 bourbon (21% rye). Bottled and released in 2015. Limited to 6000 bottles.

My Score

Notes: ______________________________

____

____

Appearance (1-5) ____ Nose (1-25) ____
Taste (1-25) ____ Finish (1-25) ____
Complexity (1-10) ____ Overall Impression (1-10) ____

SCORE ____

Middle West Michelone Reserve

Proof: 95
Age: NAS
Type: Straight
Style: Four Grain
Mash Bill: Sweet yellow corn, red winter wheat, dark pumpernickel rye, 2-row malted barley
Color: Light Amber
Price: $$$

Mike

Nose: Caramel, chocolate, corn, oak.
Taste: Caramel, tangerine, allspice, oak.
Finish: Medium long with oak and spice.

Susan

Nose: Candies fruit, vanilla icing, light baking spices, a touch of burnt sugar.
Taste: Dark fruit and caramel, baking spice, pepper, light oak.
Finish: Medium long, drying to oak tannins.

Notes

Middle West Spirits, Columbus, Ohio. Unusual choice to use sweet corn. Most bourbon is made with Dent yellow.

My Score

Notes: ______________________________

____

____

Appearance (1-5) ____ Nose (1-25) ____
Taste (1-25) ____ Finish (1-25) ____
Complexity (1-10) ____ Overall Impression (1-10) ____

SCORE ____

Milam & Greene Distillery Edition, Batch 2

Proof:	115
Age:	3 - 3 ½ years
Type:	Straight
Style:	Non-Chill Filtered, Cask Strength, Sweet Mash
Mash Bill:	70% corn, 22%rye, 8% malted barley
Color:	Dark Straw
Price:	$$

Mike

Nose: Caramel, a hint of fruit. Spice and oak.

Taste: Caramel and apples with cardamon, allspice, and some oak.

Finish: Medium long with oak and spice with a little lingering apple.

Susan

Nose: Whiff of caramel-coated chocolate with some fruit and nuts.

Taste: Caramel, chocolate, and nuts joined by some marshmallow, cinnamon, nutmeg. And apple.

Finish: Medium long with sweet oak.

Notes

Water bring out chocolate, more apple, and berries. Master Distiller Marlene Holmes crafts this. Look for more recently released batches, too.

My Score

Notes: ______________________________

____

____

Appearance (1-5) ____ Nose (1-25) ____

Taste (1-25) ____ Finish (1-25) ____

Complexity (1-10) ____ Overall Impression (1-10) ____

SCORE ____

Milam & Greene Triple Cask

Proof: 94
Age: NAS, blend of 2-year old through 11-year-old
Type: Straight
Style: Blend of straight bourbons
Mash Bill: See Distillery Edition plus Kentucky and Tennessee whiskeys
Color: Light Straw
Price: $$

Mike

Nose: Caramel apples with a hint of honey and oak.
Taste: Corn and vanilla with baking spices – cardamon and allspice – and a little apple fruit and oak.
Finish: Long with oak and spice and a little vanilla sweetness.

Susan

Nose: Caramel, pinch of herbs, citrus, corn, honey.
Taste: Caramel and dark cherries. Some tobacco and leather, too.
Finish: Spice at first leading into a soft, vanilla-touched ending.

Notes

Milam & Greene Distillery, Blanco, Texas. Master Blender Heather Greene creates the expressions that are combinations of M&G whiskey and those sourced from Kentucky and Tennessee. Look for a Single Barrel expression, too.

My Score

Notes: ____________________________________

____

____

Appearance (1-5) ____ Nose (1-25) ____
Taste (1-25) ____ Finish (1-25) ____
Complexity (1-10) ____ Overall Impression (1-10) ____

SCORE ____

Monk's Road

Proof: 104
Age: Minimum 6 years
Type: Straight
Style: Single Barrel
Mash Bill: Not available
Color: Light Straw
Price: $$$

Mike

Nose: Vanilla and leather, plus ripe apple and oak.
Taste: Vanilla and apple with a little cinnamon spice and oak.
Finish: Medium long with oak and cinnamon.

Susan

Nose: Brown sugar and banana with some roasted corn hovering in the background.
Taste: Bananas Foster with some caramel corn.
Finish: Dries to oak and some spice.

Notes

Log Still Distillery, New Haven, Kentucky. This whiskey is sourced until current distillate has aged.

My Score

Notes: ____________________________________
____
____

Appearance (1-5) ____ Nose (1-25) ____
Taste (1-25) ____ Finish (1-25) ____
Complexity (1-10) ____ Overall Impression (1-10) ____
SCORE ____

The Morning Glory

Proof: 100
Age: 4 years
Type: Straight
Style: Five Grain, Bonded
Mash Bill: 66% corn, 12% rye, 16% kasha, 2% buckwheat, 4% malted barley
Color: Light Straw
Price: $$$

Mike

Nose: Brown sugar, apples, baking spice, and oak.
Taste: Caramel, very ripe apples, cinnamon and ginger with oak and a hint of dark chocolate.
Finish: Long with oak and cinnamon and a hint of chocolate.

Susan

Nose: Apple peel, light brown sugar, some nutmeg and vanilla.
Taste: Candies nuts, apple, cinnamon, some light citrus.
Finish: Long and warm. Oaky.

Notes

Spirits of French Lick, West Baden, Indiana. An usual, but quite successful, mash bill.

My Score

Notes: ______________________________
____
____

Appearance (1-5) ____ Nose (1-25) ____
Taste (1-25) ____ Finish (1-25) ____
Complexity (1-10) ____ Overall Impression (1-10) ____
SCORE ____

Neeley Family Distillery Bourbon

Proof: 106.5
Age: 28 months
Type: Straight
Style: Pot Distilled, Small Batch
Mash Bill: 67% corn, 25% rye, 18% malted barley
Color: Dark Straw
Price: $$

Mike

Nose: Caramel, apricots, baking spice, oak.
Taste: Very fruity – apricots and peaches – very brandy-like. Nutmeg, allspice, oak, hint of leather.
Finish: Very long with oak and spice and lingering fruit and caramel.

Susan

Nose: Maple syrup dribbled caramel with crushed nuts, dried apricots.
Taste: Very nutty with fruit, brown sugar, new leather, and pepper.
Finish: Long and peppery but not hot.

Notes

Sold only at the Neeley Family Distillery in Sparta, Kentucky. Distiller Royce Neeley uses 30 gallon barrels for aging.

My Score

Notes: ______________________________

____

____

Appearance (1-5) ____ Nose (1-25) ____
Taste (1-25) ____ Finish (1-25) ____
Complexity (1-10) ____ Overall Impression (1-10) ____

SCORE ____

Neeley Family Distillery Single Barrel

Proof: 100.6
Age: 19 months
Type: Not straight
Style: Single Barrel, Wheated, Pot Distilled, Unfiltered
Mash Bill: 64% corn, 28% wheat 8% malted barley
Color: Straw
Price: $$

Mike

Nose: Caramel with a hint of milk chocolate, corn, plums, apricots, hint of oak.
Taste: Corn with lots of fruit- blackberries, raspberries, plums – a little baking spice and oak.
Finish: Medium long with very brandy-like fruit and oak.

Susan

Nose: Corn, burnt toffee, pecans, brown sugar.
Taste: Very rich mouthfeel with vanilla, pecans, and fruit salad.
Finish: Fresh corn husk to start, drying to pepper and oak.

Notes

Neeley Family Distillery, Sparta, Kentucky. Aged in 25 gallon barrels.

My Score

Notes: ____________________________________
____
____

Appearance (1-5) ____ Nose (1-25) ____
Taste (1-25) ____ Finish (1-25) ____
Complexity (1-10) ____ Overall Impression (1-10) ____
SCORE ____

Never Say Die

Proof: 95
Age: NAS
Type: Straight
Style: Small Batch
Mash Bill: 75% corn, 21% rye, 4% malted barley
Color: Straw
Price: $$$

Mike

Nose: Vanilla, apples, peaches, oak.
Taste: Vanilla, apples, red pepper, oak. Thin mouthfeel.
Finish: Long with oak and red pepper.

Susan

Nose: Vanilla, orange zest, honey, savory spice.
Taste: Brown sugar, then fruit yielding to pepper.
Finish: Dry with a dash of pepper flakes.

Notes

Manifest Destiny Distilling, Jacksonville, Florida and NSD Spirits, Great Britain. Sourced from an undisclosed Kentucky distillery and aged and bottled in Great Britain. Named for an underdog champion racehorse.

My Score

Notes: ____

____

____

Appearance (1-5) ____ Nose (1-25) ____
Taste (1-25) ____ Finish (1-25) ____
Complexity (1-10) ____ Overall Impression (1-10) ____

SCORE ____

Neversweat

Proof: 80
Age: NAS
Type: Straight
Style: Craft
Mash Bill: Not available
Color: Light Straw
Price: $$

Mike

Nose: Corn and vanilla with a hint of oak. Very light and simple.
Taste: Very young with sweet vanilla and corn. Some fruit. Peaches? Apricots? Pepper spice.
Finish: Very good. Starts sweet but dries out with some oak tannins and peppery spice.

Susan

Nose: Spicy – anise and clove.
Taste: Sweet spices with caramel and vanilla arriving next with some fruit. Tannins arrive last.
Finish: Quite smooth for a bourbon with such a spicy start. Bit of cinnamon aftertaste.

Notes

Made by Headframe Spirits of Butte, Montana, which also makes craft gin and vodka. Beverage names, as well as the distillery name, come from mining terms.

My Score

Notes: ______________________________________

____

____

Appearance (1-5) ____ Nose (1-25) ____
Taste (1-25) ____ Finish (1-25) ____
Complexity (1-10) ____ Overall Impression (1-10) ____

SCORE ____

New Riff Bottled-in-Bond

Proof: 100
Age: 4 years
Type: Straight
Style: Bonded
Mash Bill: 65% corn, 30% rye, 5% malted barley
Color: Straw
Price: $$$

Mike

Nose: French vanilla, sorn, some oak, pitted fruit.
Taste: Vanilla, corn, oak, and cherries.
Finish: Medium long with oak and spice.

Susan

Nose: Vanilla, caramel dusting of nutmeg.
Taste: Caramel, vanilla, ripe apples, cinnamon.
Finish: Very nice. Long and dry with oak but retains some vanilla.

Notes

New Riff Distilling, Newport, Kentucky. Equally excellent for sipping of cocktails.

My Score

Notes: ______________________________

____

____

Appearance (1-5) ____ Nose (1-25) ____
Taste (1-25) ____ Finish (1-25) ____
Complexity (1-10) ____ Overall Impression (1-10) ____

SCORE ____

New Southern Revival Four Grain

Proof: 94
Age: NAS
Type: Not Straight
Style: Wheated
Mash Bill: See notes
Color: Light Straw
Price: $$$

Mike

Nose: Young. Corn with hint of honey.
Taste: Young with honey and spice. Hint of peaches.
Finish: Short and sweet.

Susan

Nose: Sweet corn. Simple.
Taste: Lots of grain. Some vanilla.
Finish: Short. Oaky at the end.

Notes

Made by High Wire Distilling Co., Charleston, South Carolina. Mash bill uses heirloom white corn, red winter wheat, Carolina Gold rice, and malted barely. This was bottle 187 of batch 3.

My Score

Notes: ______________________________

____

____

Appearance (1-5) ____ Nose (1-25) ____
Taste (1-25) ____ Finish (1-25) ____
Complexity (1-10) ____ Overall Impression (1-10) ____
SCORE ____

Oak Eye

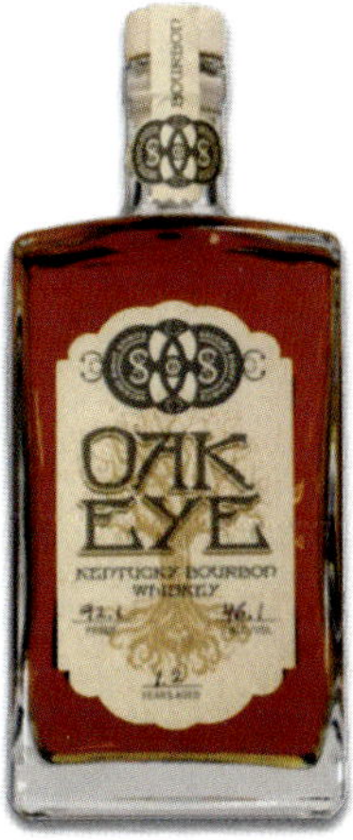

Proof: 92.2
Age: 2.2 years
Type: Straight
Style: Small Batch, Four Grain
Mash Bill: 72% corn, 5% rye, 16% wheat, 7% malted barley
Color: Dark Straw
Price: $$

Mike

Nose: Vanilla, hint of fruit spice, oak.
Taste: Vanilla, apricots, black pepper, oak.
Finish: Medium long with oak and pepper.

Susan

Nose: Brown sugar, sweet oak, applesauce.
Taste: Brown sugar, oak, citrus.
Finish: Medium long and well balanced.

Notes

Very, very good for a young bourbon. Second Sight Spirits, Ludlow, Kentucky. Proof and age vary with batches. Low barrel entry proof of 115.

My Score

Notes: ____

____

____

Appearance (1-5) ____ Nose (1-25) ____
Taste (1-25) ____ Finish (1-25) ____
Complexity (1-10) ____ Overall Impression (1-10) ____

SCORE ____

Oaklore Four Grain

Proof: 92
Age: NAS
Type: Straight
Style: Small Batch, Four Grain
Mash Bill: 74% corn, 5% rye, 11% wheat, 10% malted barley
Color: Straw
Price: $$$

Mike

Nose: Vanilla, apples, baking spice, oak.
Taste: Vanilla, apples, pears cinnamon, oak.
Finish: Medium long with oak and cinnamon.

Susan

Nose: Vanilla, oak, medley of baking spices.
Taste: Dark brown sugar, nuts, apples, cinnamon.
Finish: Dries quickly to spice oak.

Notes

Oaklore Distilling Co., Matthews, North Carolina. Distilled in Kentucky and North Carolina.

My Score

Notes: ____________________________________

____

____

Appearance (1-5) ____ Nose (1-25) ____
Taste (1-25) ____ Finish (1-25) ____
Complexity (1-10) ____ Overall Impression (1-10) ____

SCORE ____

OCD #5

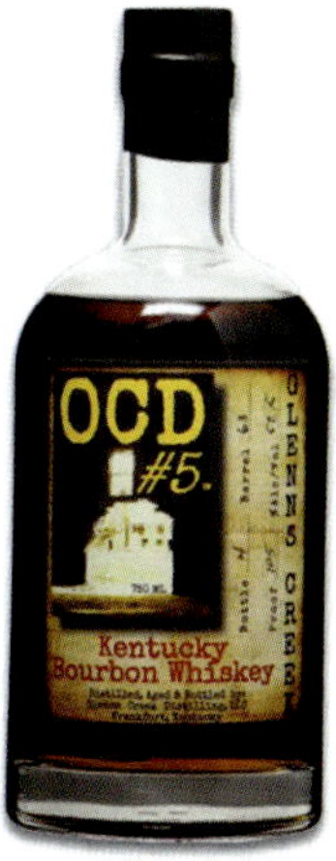

Proof: 92
Age: NAS
Type: Straight
Style: Barrel Strength
Mash Bill: 87% corn, 8% rye, 5% malted barley
Color: Dark Straw
Price: $$$

Mike

Nose: Vanilla, chocolate, apples, oak.
Taste: Vanilla, milk chocolate, apples, cinnamon, oak.
Finish: Medium long with oak, chocolate, and lingering apple.

Susan

Nose: Multigrain baking bread with some raisins.
Taste: Lightly smokey, dark cherry, nuts, dark brown sugar, tobacco leaf.
Finish: Dry with a brown sugar note at the end.

Notes

Glenns Creek Distilling, Frankfort, Kentucky. Located in the former Old Crow Distillery.

My Score

Notes: ______________________________

____

____

Appearance (1-5) ____ Nose (1-25) ____
Taste (1-25) ____ Finish (1-25) ____
Complexity (1-10) ____ Overall Impression (1-10) ____

SCORE ____

Off Hours

Proof: 95
Age: 5 years
Type: Straight
Style: Traditional
Mash Bill: 75% corn, 21% rye, 4% malted barley
Color: Light Straw
Price: $$

Mike

Nose: Vanilla and hazelnuts with a little oak and spice.
Taste: Vanilla, apples, hint of pecan, oak, baking spices.
Finish: Medium long with oak and spice and a little lingering pecan.

Susan

Nose: Bit-O-Honey candy. Sweet and nutty, some light citrus and cherries.
Taste: Vanilla custard with fruit, nuts. and honey sprinkled with light brown sugar.
Finish: Long and warm.

Notes

Nicely balanced. Made at MGP in Lawrenceburg, Indiana and bottled in Bloomington.

My Score

Notes: ____

____

____

Appearance (1-5) ____ Nose (1-25) ____
Taste (1-25) ____ Finish (1-25) ____
Complexity (1-10) ____ Overall Impression (1-10) ____
SCORE ____

O.H. Ingram River Aged

Proof: 105
Age: NAS
Type: Straight
Style: Wheated
Mash Bill: Corn, wheat, malted barley, but no proportions disclosed
Color: Straw
Price: $$$

Mike

Nose: Caramel, a little baking spice and banana with a hint of oak.
Taste: Caramel with a little banana and cinnamon. Some sweet oak.
Finish: Long with sweet oak and spice.

Susan

Nose: Canned peaches, vanilla wafers, some light brown sugar.
Taste: Matches the nose with more vanilla. Apples join the peaches.
Finish: Fruit persists and is joined by pepper and tannins.

Notes

Brown Water Spirits, Owensboro, Kentucky. Sourced and then aged on a barge moored on the banks of the Mississippi River in Ballard County, Kentucky.

My Score

Notes: ________________________________

____

____

Appearance (1-5) ____ Nose (1-25) ____
Taste (1-25) ____ Finish (1-25) ____
Complexity (1-10) ____ Overall Impression (1-10) ____

SCORE ____

O.H. Ingram River Aged Flagship

Proof: 100
Age: NAS
Type: Straight
Style: Small Batch
Mash Bill: Undisclosed
Color: Light Straw
Price: $$$

Mike

Nose: Very light. Vanilla with a hint of fruit and oak.
Taste: Corn, vanilla, light berry fruit – raspberry and blackberry. Some baking spice and oak.
Finish: Medium long with oak and spice.

Susan

Nose: Vanilla with a little horehound candy and baking spice.
Taste: Fruity with ripe apples, plus caramel, some pears, and baking spice.
Finish: Sweetness fades to peppery oak.

Notes

Brown Water Spirits, Owensboro, Kentucky. Sourced from MGP, then aged in a floating barge moored on the Mississippi River in Ballard County, Kentucky.

My Score

Notes: ______________________________

____

____

Appearance (1-5) ____ Nose (1-25) ____
Taste (1-25) ____ Finish (1-25) ____
Complexity (1-10) ____ Overall Impression (1-10) ____

SCORE ____

Old 55 Bourbon

Proof:	80
Age:	NAS
Type:	Not Straight
Style:	Single Barrel
Mash Bill:	100% sweet corn
Color:	Very Light Straw
Price:	$$$

Mike

Nose: Sweet and floral with banana and vanilla. Banana pudding.

Taste: Corn and vanilla with banana and a hint of oak and spice.

Finish: Very sweet with a hint of oak and spice.

Susan

Nose: Cornbread with a whiff of sweet mint and very ripe apples.

Taste: Sweet corn, ripe apples and light caramel.

Finish: Sweet, with a bit of pepper and a distinct floral note at the end.

Notes

From Old 55 Distillery, Newtown, Indiana. Produces "estate whiskeys" by growing, milling, and mashing their own grains. Aged in 30 gallon barrels.

My Score

Notes: ______________________________

____

____

Appearance (1-5)	____	Nose (1-25)	____
Taste (1-25)	____	Finish (1-25)	____
Complexity (1-10)	____	Overall Impression (1-10)	____

SCORE ____

Old 55 Single Barrel

Proof:	80
Age:	NAS
Type:	Not Straight
Style:	Single Barrel
Mash Bill:	Not available
Color:	Very Light Straw
Price:	$$$

Mike

Nose: Corn and vanilla with some leather.
Taste: Buttered corn and pepper spice with some oak.
Finish: Dry with oak and pepper. A little creamy.

Susan

Nose: Corn husk, tobacco leaf and tobacco smoke.
Taste: Smoky, like a mouthful of pipe tobacco smoke. Some nuttiness.
Finish: Lingering and dry with oak tannins.

Notes

From Old 55 Distillery, Newtown, Indiana. Produces "estate whiskeys" by growing, milling, and mashing their own grains. Aged in 30 gallon barrels. Batch 15H5, bottle 108.

My Score

Notes: ______________________________________

____

____

Appearance (1-5) ____ Nose (1-25) ____
Taste (1-25) ____ Finish (1-25) ____
Complexity (1-10) ____ Overall Impression (1-10) ____

SCORE ____

Old Dominick Huling Station

Proof: 100
Age: NAS
Type: Straight
Style: Traditional
Mash Bill: 52% corn, 44% rye, 4% malted barley
Color: Straw
Price: $$

Mike

Nose: Vanilla and citrus with some rose and honeysuckle.
Taste: Vanilla, roses, baking spice, and oak. Rich mouthfeel.
Finish: Medium long with oak, vanilla and some baking spices.

Susan

Nose: Vanilla, cherries, and a little baking spice along with floral notes.
Taste: Vanilla, a dash of cinnamon, caramel and a bit of tobacco.
Finish: Medium long with some subtle chocolate.

Notes

Old Dominick Distillery, Memphis, Tennessee. A little water brings out a fruit medley.

My Score

Notes: ______________________________

____

____

Appearance (1-5) ____ Nose (1-25) ____
Taste (1-25) ____ Finish (1-25) ____
Complexity (1-10) ____ Overall Impression (1-10) ____

SCORE ____

Old Elk Blended Straight Bourbon

Proof:	88
Age:	NAS
Type:	Straight
Style:	Blend
Mash Bill:	51% corn, 15% rye, 34% malted barley
Color:	Straw
Price:	$$

Mike

Nose: Very light. A little vanilla and corn with a hint of pears.

Taste: Thin mouthfeel. Caramel, pears, pepper, and a hint of oak.

Finish: Long and peppery with only a hint of oak.

Susan

Nose: Apple and vanilla with some pecans and a bit of cinnamon.

Taste: Vanilla with some cinnamon, orchard fruit and light oak.

Finish: Peppery, smoothing to fruit.

Notes

Old Elk Distillery, Fort Collins, Colorado. Other expressions include annual Master's Blend series, and a wheated bourbon.

My Score

Notes: ______________________________

____

____

Appearance (1-5) ____ Nose (1-25) ____

Taste (1-25) ____ Finish (1-25) ____

Complexity (1-10) ____ Overall Impression (1-10) ____

SCORE ____

Old Elk Single Barrel

Proof: 118.1
Age: 5 years
Type: Straight
Style: Wheated, Cask Strength
Mash Bill: 51% corn, 45% wheat, 4% malted barley
Color: Light Amber
Price: $$$

Mike

Nose: Caramel, apples, hint of almonds and chocolate, with white pepper and oak.
Taste: Caramel apples with white pepper and oak.
Finish: Long with oak, pepper, and a bit of chocolate.

Susan

Nose: Very caramel, with book leather, baking spice, and dark fruit.
Taste: Caramel, dark cherries, cinnamon, nutmeg, ginger. Rich mouthfeel.
Finish: Long and complex.

Notes

This was a private barrel selection for Justins' House of Bourbon. Five drops of water amplify the apple and brings out the chocolate.

My Score

Notes: ______________________________

____

____

Appearance (1-5) ____ Nose (1-25) ____
Taste (1-25) ____ Finish (1-25) ____
Complexity (1-10) ____ Overall Impression (1-10) ____

SCORE ____

Old Forge Reserve

Proof: 89
Age: 9 years
Type: Not Straight
Style: Single Barrel
Mash Bill: Not available
Color: Straw
Price: $$$

Mike

Nose: Maple smoke with vanilla and buttered corn.
Taste: Maple syrup, corn, vanilla, and oak.
Finish: Long lasting maple smoke.

Susan

Nose: Maple, corn, some cinnamon.
Taste: Sweet apples, sweet oak, flash of maple and smoke.
Finish: Warm, smooth, smoky.

Notes

A Tennessee bourbon (probably from George Dickel) bottled by Old Forge Distillery, Pigeon Forge, Tennessee. Barrel #16-773.

My Score

Notes: ____________________

____

____

Appearance (1-5) ____ Nose (1-25) ____
Taste (1-25) ____ Finish (1-25) ____
Complexity (1-10) ____ Overall Impression (1-10) ____

SCORE ____

Old Fourth Distillery Bottled-in-Bond

Proof: 100
Age: 4 years
Type: Straight
Style: Bonded, Single Barrel
Mash Bill: 51% corn, 45% rye, 4% malted barley
Color: Straw
Price: $$$

Mike

Nose: Caramel and a hint of milk chocolate, with a little ripe apple or pear, plus oak.
Taste: Caramel and chocolate with a little pear, white pepper, and oak.
Finish: Long with oak, white pepper, and lingering caramel.

Susan

Nose: Vanilla, some fruit and spice, light brown sugar, and a whiff of cocoa.
Taste: Apple peel, figs, dark brown sugar, cinnamon, and a touch of book leather.
Finish: Long and spicy.

Notes

The Old Fourth Distillery in Atlanta, Georgia closed in January 2023. However, the bourbon may still be encountered in well-stocked whiskey bars. Bottle 92 of barrel 54.

My Score

Notes: ______________________________

____

____

Appearance (1-5) ____ Nose (1-25) ____
Taste (1-25) ____ Finish (1-25) ____
Complexity (1-10) ____ Overall Impression (1-10) ____

SCORE ____

Old Hell Roaring

Proof: 90
Age: NAS
Type: Straight
Style: Double Barreled
Mash Bill: Not available
Color: Dark Straw
Price: $$$

Mike

Nose: Bananas, caramel, and a little oak.
Taste: Oak tannins, vanilla, and bananas.
Finish: Dry with lots of oak.

Susan

Nose: Pear, vanilla, and a bit of smoke.
Taste: Vanilla, baking spice, pears.
Finish: Some smoke and spice with a floral note.

Notes

Crooked Water Spirits, Minneapolis, Minnesota, Double barreled. Barrels specially toasted and smoked.

My Score

Notes: ______________________________

____

____

Appearance (1-5) ____ Nose (1-25) ____
Taste (1-25) ____ Finish (1-25) ____
Complexity (1-10) ____ Overall Impression (1-10) ____
SCORE ____

Old Kentucky

Proof:	80
Age:	4 years
Type:	Straight
Style:	Traditional
Mash Bill:	Not available
Color:	Straw
Price:	$$

Mike

Nose: Light. Very young. Some vanilla and and corn.
Taste: Thin and watery. Some vanilla and corn, a hint of oak. Not much else.
Finish: Short and slightly oaky, but disappointing.

Susan

Nose: Light. Some mint. More tobacco. A little vanilla as it sits.
Taste: Roasted corn and some candied nuts. Simple, but pleasant at 80 proof.
Finish: Smooth, no burn. But short.

Notes

Bottled by Distiller Products, LTD., San Jose, California. Sourced from Kentucky, but no indication from whom.

My Score

Notes: ______________________________

____

____

Appearance (1-5) ____ Nose (1-25) ____
Taste (1-25) ____ Finish (1-25) ____
Complexity (1-10) ____ Overall Impression (1-10) ____

SCORE ____

Old Log Cabin

Proof: 86
Age: At least 18 months
Type: Not Straight
Style: Two Grain
Mash Bill: 51% corn, 49% malted barley
Color: Light Straw
Price: $$$

Mike

Nose: Corn and malt, not surprisingly with a hint of vanilla.
Taste: Corn and vanilla with some malty notes.
Finish: Corn and a hint of pepper and wood.

Susan

Nose: Breakfast cereal. Some honey sweetness.
Taste: Barley malt with some corn sweetness. Very like a Scotch. perhaps a pear note.
Finish: Short and dry.

Notes

We loved the cabin-shaped bottle. Very interesting to taste what happens when there is not rye or wheat. Comes from Batch 206 Distillery, Seattle, Washington.

My Score

Notes: ______________________________

____

____

Appearance (1-5) ____ Nose (1-25) ____
Taste (1-25) ____ Finish (1-25) ____
Complexity (1-10) ____ Overall Impression (1-10) ____

SCORE ____

Old Louisville Whiskey Co. Bourbon

Proof: 114
Age: NAS
Type: Straight
Style: Cask Strength
Mash Bill: 75% corn, 21% rye, 4% malted barley
Color: Pale Straw
Price: $$$$

Mike

Nose: Vanilla, corn, pears, oak.
Taste: Vanilla, corn, pears, white pepper, oak.
Finish: Medium long with oak and pepper.

Susan

Nose: Pears, vanilla icing, and some baking spice.
Taste: Fruit forward with vanilla and some baking spice.
Finish: Long, warm, smooth.

Notes

Old Louisville Whiskey Co., Louisville, Kentucky. Website states this was aged a minimum of 7 years.

My Score

Notes: ______________________________

____

____

Appearance (1-5) ____ Nose (1-25) ____
Taste (1-25) ____ Finish (1-25) ____
Complexity (1-10) ____ Overall Impression (1-10) ____

SCORE ____

Old McBrayer Bourbon Bottled-in-Bond

Proof:	100
Age:	NAS, at least 4 years
Type:	Straight
Style:	Bonded
Mash Bill:	70% corn, 18% rye, 12% malted barley
Color:	Dark Straw
Price:	$$$

Mike

Nose: Vanilla, caramel, pears, baking spice, oak.
Taste: Vanilla, pears, cinnamon, nutmeg, oak.
Finish: Medium long with oak and cinnamon.

Susan

Nose: Peach pie crust and brown sugar = peach cobbler.
Taste: Vanilla crème brulee, pears, cinnamon, touch of oak.
Finish: Long and balanced.

Notes

McBrayer Legacy Spirits, Bardstown, Kentucky. William Harrison McBrayer of Lawrenceburg was a 19th century judge, farmer, distiller, senator, and philanthropist.

My Score

Notes: ______________________________

____

____

Appearance (1-5) ____ Nose (1-25) ____
Taste (1-25) ____ Finish (1-25) ____
Complexity (1-10) ____ Overall Impression (1-10) ____

SCORE ____

Old Medley

Proof: 86.8
Age: 12 years
Type: Straight
Style: Extra Aged
Mash Bill: Proprietary
Color: Dark Straw
Price: $$$

Mike

Nose: Caramel and apples with some fine leather and oak.

Taste: Caramel and fruit. Apples and pears with some pepper spice on the end. Nice mouthfeel. No burn and some warmth in the throat.

Finish: Long and dry with oak tannins balanced by a bit of sweet caramel.

Susan

Nose: Lovely honey nose with caramel and some cherries.

Taste: Vanilla, pecans, a little new leather, some orange peel and black pepper. Pepper definitely in the background.

Finish: Fruit gives way to soft oak. Long and warm with no burn at all.

Notes

Very elegant sip. Kentucky-sourced. Distributed by the historic Charles Medley Distillery, Owensboro, Kentucky, which is in the process of beginning to make bourbons again.

My Score

Notes: ____________________________________

____

____

Appearance (1-5) ____ Nose (1-25) ____
Taste (1-25) ____ Finish (1-25) ____
Complexity (1-10) ____ Overall Impression (1-10) ____

SCORE ____

Old Pepper

Proof: 117.8
Age: 10 years
Type: Straight
Style: Cask Strength
Mash Bill: Not available
Color: Light Straw
Price: $$$$

Mike

Nose: Vanilla, citrus, leather, and oak with a hint of spice.
Taste: Vanilla, orange zest, leather tobacco, ginger, and nutmeg with a little oak.
Finish: Lots of oak and ginger spice.

Susan

Nose: Peach, vanilla, rising dough.
Taste: Sweet grass and peaches with some vanilla and oak.
Finish: Faintly oaky.

Notes

James E. Pepper Distilling Co., Lexington, Kentucky. This bourbon was sourced from undisclosed distillery. Water opens it up for much more complexity.

My Score

Notes: ______________________________________

____

____

Appearance (1-5) ____ Nose (1-25) ____
Taste (1-25) ____ Finish (1-25) ____
Complexity (1-10) ____ Overall Impression (1-10) ____

SCORE ____

Old Pogue

Proof: 91
Age: NAS
Type: Straight
Style: Craft
Mash Bill: Not available
Color: Dark Straw
Price: $$$

Mike

Nose: Caramel corn and oak. very simple nose. Bit fruit (pears?) and sweet spice.
Taste: Corn and vanilla with a bit of spice and tobacco.
Finish: Starts sweet, then gets spicy and dry. Lasts fairly long and gets drier and more oaky as it lasts.

Susan

Nose: Sweet buttered corn, caramel, dark fruit (plum? black cherry?)
Taste: Light caramel, buttered popcorn, hint of sweet maple syrup.
Finish: Long and smooth. Stays sweet until it dries to an oaky ending.

Notes

A fine sipping bourbon with enough presence and proof to make excellent cocktails, such as a classic Manhattan. Use top shelf vermouth.

My Score

Notes: ______________________________

____

____

Appearance (1-5) ____ Nose (1-25) ____
Taste (1-25) ____ Finish (1-25) ____
Complexity (1-10) ____ Overall Impression (1-10) ____

SCORE ____

Old Soul Bourbon

Proof: 90
Age: 16 months
Type: Not straight
Style: Non-Chill Filtered
Mash Bill: High rye
Color: Light Amber
Price: $$

Mike

Nose: Vanilla with a hint of peaches and oak.
Taste: Corn husk, vanilla, peaches, and oak.
Finish: Medium long, with oak and pepper.

Susan

Nose: Corn, some oak and vanilla.
Taste: Caramel corn with light pepper and hint of fruit.
Finish: Peppery.

Notes

Cathead Distillery, Jackson, Mississippi. Blend of sourced MGP and Cathead's bourbons. At least one mash bill 21% rye. Released February 2021. More recent releases exist.

My Score

Notes: ____

____

____

Appearance (1-5) ____ Nose (1-25) ____
Taste (1-25) ____ Finish (1-25) ____
Complexity (1-10) ____ Overall Impression (1-10) ____

SCORE ____

Old Williamsburg

Proof: 80
Age: 36 months
Type: Straight
Style: Traditional
Mash Bill: Not available
Color: Extremely Pale Straw
Price: $

Mike

Nose: Very light, some corn and vanilla.
Taste: Very light. Some sweet corn and vanilla. Brown vodka.
Finish: Short and sweet.

Susan

Nose: Almost nonexistent.
Taste: Astringent! A little corn.
Finish: Mercifully short.

Notes

Old Williamsburg Products Co., Princeton, Minnesota. Ginger ale has more color than this stuff. Sourced from Kentucky. (Some distiller's rejects?)

My Score

Notes: ______________________________

____

____

Appearance (1-5) ____ Nose (1-25) ____
Taste (1-25) ____ Finish (1-25) ____
Complexity (1-10) ____ Overall Impression (1-10) ____
SCORE ____

Old Williamsburg No. 20

Proof:	101
Age:	36 months
Type:	Straight
Style:	Traditional
Mash Bill:	Not available
Color:	Very Pale Straw
Price:	$$

Mike

Nose: Corn and vanilla with just a hint of flowers.
Taste: Corn and vanilla with some pepper spice.
Finish: Medium long and spicy.

Susan

Nose: Faint vanilla, carnations. Very light.
Taste: Some vanilla and oak, but very.
Finish: Medium long, hot.

Notes

Old Williamsburg Products Co., Princeton, Minnesota. Sourced from Kentucky. (Really?)

My Score

Notes: ______________________________

____

____

Appearance (1-5) ____ Nose (1-25) ____
Taste (1-25) ____ Finish (1-25) ____
Complexity (1-10) ____ Overall Impression (1-10) ____
SCORE ____

Oppidan

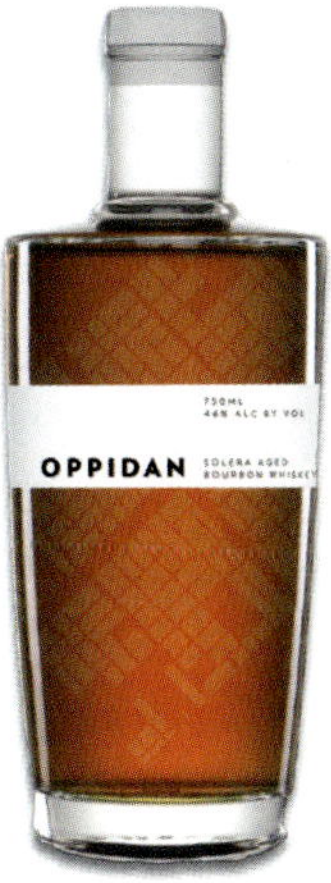

Proof:	92
Age:	NAS
Type:	Not Straight
Style:	Special Finish
Mash Bill:	See notes
Color:	Light Amber
Price:	$$

Mike

Nose: Bread dough and caramel apple.
Taste: Caramel and apple with some leather and spice.
Finish: Dry with oak and pepper spice.

Susan

Nose: Corn, oak, then caramel.
Taste: Sweet corn, some fig, baking spices.
Finish: Herbal at first. Finishes with sweet corn.

Notes

From Oppidan Spirits, Wheeling, Illinois. Not complex, but nicely made. Distillers use the Solara process. The 5-grain mash bill contains corn, malted rye, chocolate malted rye, special B and 2-row barley.

My Score

Notes: ______________________________

____

____

Appearance (1-5) ____ Nose (1-25) ____
Taste (1-25) ____ Finish (1-25) ____
Complexity (1-10) ____ Overall Impression (1-10) ____
SCORE ____

OYO Bourbon Whiskey Michelone Reserve

Proof: 90
Age: NAS
Type: Not Straight
Style: See notes
Mash Bill: See notes
Color: Dark Straw
Price: $$

Mike

Nose: Very light. Vanilla and honey. Water brings out fruit.
Taste: Vanilla and peaches with some nutmeg and cinnamon. Water brings out vanilla.
Finish: Long and dry with pepper spice and oak. Water sweetens the finish.

Susan

Nose: Candy corn, sweet spices. Very light. Water amplifies alcohol.
Taste: Initially sweet corn and fruit, but quickly becomes peppery hot.
Finish: Long and hot. Water tames heat resulting in a nutty finish.

Notes

Midwest Spirits, LLC, Columbus, Ohio. This is a mingling of a Kentucky sourced rye recipe bourbon with a wheat whiskey made at the distillery. While it says "bourbon" on the label, by law this cannot be called bourbon because something, namely wheat whiskey, has been added to it. Interesting whiskey, but not bourbon. We have included it to illustrate the distinction.

My Score

Notes: ____

____

____

Appearance (1-5) ____ Nose (1-25) ____
Taste (1-25) ____ Finish (1-25) ____
Complexity (1-10) ____ Overall Impression (1-10) ____

SCORE ____

Paddleford Creek

Proof: 93
Age: NAS
Type: Not Straight
Style: Small Batch
Mash Bill: Not available
Color: Almost None
Price: $$

Mike

Nose: Corn and vanilla, but very light.
Taste: Thin and watery. Beige vodka.
Finish: Hint of vanilla and oak. Most of its flavor is in the finish.

Susan

Nose: Very, very faint.
Taste: Some pear and vanilla, but very light.
Finish: Very short.

Notes

Produced and bottled by Paddleford Spirits, Princeton, Minnesota. No indication of source.

My Score

Notes: ______________________________

____

____

Appearance (1-5) ____ Nose (1-25) ____
Taste (1-25) ____ Finish (1-25) ____
Complexity (1-10) ____ Overall Impression (1-10) ____

SCORE ____

Peerless Batch 2

Proof: 109.5
Age: 4 years
Type: Straight
Style: Sweet Mash, Small Batch, Cask Strength
Mash Bill: Not available
Color: Light Amber
Price: $$$

Mike

Nose: Caramel with a hint of chocolate, ripe apples and spice. Hint of oak.
Taste: Caramel apples, cinnamon, nutmeg, and a little oak.
Finish: Long with oak, milk chocolate, and cinnamon.

Susan

Nose: Crème brulee and gingerbread with some berries and cocoa.
Taste: Freshly sliced apple, brown sugar, vanilla, and baking spices.
Finish: Lingers, with some chocolate.

Notes

Kentucky Peerless Distilling Co., Louisville, Kentucky. Low barrel entry proof brings out caramel and chocolate. One of several limited releases that the distillery offers several times a year. Look for others, too.

My Score

Notes: ______________________________

____

____

Appearance (1-5) ____ Nose (1-25) ____
Taste (1-25) ____ Finish (1-25) ____
Complexity (1-10) ____ Overall Impression (1-10) ____
SCORE ____

Peerless Small Batch

Proof: 108.9
Age: 4 years
Type: Straight
Style: Sweet Mash, Small Batch, Cask Strength
Mash Bill: Not available
Color: Light Amber
Price: $$$

Mike

Nose: Vanilla, pears, hints of spice and oak.
Taste: Vanilla, pears, cinnamon, nutmeg, oak, and a hint of tobacco.
Finish: Long and dry with oak and cinnamon.

Susan

Nose: Caramel and vanilla with oak, pears, and baking spices.
Taste: Caramel, vanilla, ripe pears, cinnamon. Water releases some milk chocolate.
Finish: Long and spicy with some figs.

Notes

Kentucky Peerless Distilling Co., Louisville, Kentucky.

My Score

Notes: ______________________________

____

____

Appearance (1-5) ____ Nose (1-25) ____
Taste (1-25) ____ Finish (1-25) ____
Complexity (1-10) ____ Overall Impression (1-10) ____

SCORE ____

Pendergast's Royal Gold

Proof:	90
Age:	5 years
Type:	Straight
Style:	Four Grain, Wheated, Finished
Mash Bill:	Corn, rye, wheat, malted barley
Color:	Dark Straw
Price:	$$$

Mike

Nose: Caramel with some leather and baking spices.
Taste: Caramel with a bit of ripe apples, baking spices, and oak.
Finish: Long and dry with oak and a hint of nutmeg.

Susan

Nose: Caramel, baked apples, hint of milk chocolate.
Taste: Caramel, sweet apples, cinnamon, chocolate.
Finish: Long and spicy with a wine note at the end.

Notes

Tom's Town Distilling Company, Kansas City, Missouri. Finished in ex-port wine casks. Well balanced.

My Score

Notes: ______________________________

____

____

Appearance (1-5) ____ Nose (1-25) ____
Taste (1-25) ____ Finish (1-25) ____
Complexity (1-10) ____ Overall Impression (1-10) ____

SCORE ____

Penny Packer

Proof: 80
Age: NAS
Type: Straight Bourbon
Style: Traditional
Mash Bill: Proprietary
Color: Straw
Price: $$

Mike

Nose: Sweet vanilla and cherry. Hint of tobacco.
Taste: Very light. Sweet vanilla and cherries, dates. A bit of pepper spice.
Finish: Starts sweet, but gets peppery and dry with oak tannins.

Susan

Nose: Sweet corn and vanilla. Light, sweet fruit.
Taste: Sweet corn and cherry. Very lightly spicy.
Finish: Medium long. Very nice for low proof. Sweet corn dries to light oak.

Notes

This is distilled by Heaven Hill in Louisville for export to Germany where it is bottled. It has now found its way back to the U.S. Distributed by Niche Import Co. Limited availability.

My Score

Notes: ______________________________

____

____

Appearance (1-5) ____ Nose (1-25) ____
Taste (1-25) ____ Finish (1-25) ____
Complexity (1-10) ____ Overall Impression (1-10) ____
SCORE ____

Pinhook Bourbon Heist

Proof: 98
Age: 3 years
Type: Straight
Style: Unfiltered
Mash Bill: 51% corn, 45% rye, 4% malted barley
Color: Light Straw
Price: $$$

Mike

Nose: Fruit forward, very brandy-like. Pears and apples with vanilla and a hint of spice and oak.
Taste: Apples and vanilla with some allspice and cinnamon, sweet oak. Not at all tannic.
Finish: Long with oak and cinnamon.

Susan

Nose: Sweet corn, a little leather, and some apple.
Taste: Roasted sweet corn, with cinnamon, apples, and a touch of milk chocolate.
Finish: Long and sweet until ending in a bit of oak and light pepper.

Notes

Pinhook Kentucky Straight Bourbon Whiskey. Distilled at Castle & Key, Frankfort, Kentucky. One of several expressions available.

My Score

Notes: ______________________________

____

____

Appearance (1-5) ____ Nose (1-25) ____
Taste (1-25) ____ Finish (1-25) ____
Complexity (1-10) ____ Overall Impression (1-10) ____
SCORE ____

Prestige

Proof: 80
Age: 36 months
Type: Straight
Style: Traditional
Mash Bill: Not available
Color: Light Straw
Price: $

Mike

Nose: Very light corn and vanilla.
Taste: Thin and watery, but some nice buttered corn and vanilla.
Finish: Short with a hint of oak.

Susan

Nose: Sweet mint, light caramel corn.
Taste: Carmel apple and sprinkling of sweet spice.
Finish: Dries to black pepper.

Notes

Sourced from Kentucky. Bottled by J. Harrison Co., Mira Loma, Californa.

My Score

Notes: ______________________________________

____

____

Appearance (1-5) ____ Nose (1-25) ____
Taste (1-25) ____ Finish (1-25) ____
Complexity (1-10) ____ Overall Impression (1-10) ____

SCORE ____

Rabbit Hole Cavehill

Proof: 95
Age: 3 years
Type: Straight
Style: Wheated, Four Grain
Mash Bill: 70% corn, 10% malted wheat, 10% malted barley, 10% honey malt
Color: Pale Straw
Price: $$$

Mike

Nose: Vanilla with a touch of honey, a hint of faint spice, and oak.
Taste: Vanilla and honey with some citrus, cinnamon, and oak.
Finish: Long with spice and oak and a little lingering honey.

Susan

Nose: Red raspberries, strawberries, some vanilla bean, and new leather.
Taste: Vanilla and bananas, joined by some citrus, chewy caramel, and a dash of cinnamon.
Finish: Medium long drying to a little pepper.

Notes

The distillery in the NuLu neighborhood of Louisville is still new. This was contract distilled at new Riff in Newport, Kentucky.

My Score

Notes: ______________________________

____

____

Appearance (1-5) ____ Nose (1-25) ____
Taste (1-25) ____ Finish (1-25) ____
Complexity (1-10) ____ Overall Impression (1-10) ____

SCORE ____

Rabbit Hole Sherry

Proof: 93
Age: NAS
Type: Straight
Style: Wheated, Special Finish
Mash Bill: Uses wheat
Color: Straw
Price: $$$

Mike

Nose: Raisins and dates with vanilla and a hint of baking spice.
Taste: Raisins and berries with caramel corn and baking spices.
Finish: Short and fruity sweet with oak and spice.

Susan

Nose: Sherry predominates. Some vanilla, dried fruit.
Taste: Really, this whiskey tastes like sherry, complete with the sweet nuttiness.
Finish: Sweet with raisins and touch of spice.

Notes

Contract distilled to the owner's specifications. No point of origin on the label, and "Kentucky" is absent. Finished in 30 year-old PX sherry casks. The company is building a distillery in downtown Louisville.

My Score

Notes: ______________________________

____

____

Appearance (1-5) ____ Nose (1-25) ____
Taste (1-25) ____ Finish (1-25) ____
Complexity (1-10) ____ Overall Impression (1-10) ____

SCORE ____

Rabbit Hole Straight Bourbon

Proof: 95
Age: NAS
Type: Straight
Style: Wheated
Mash Bill: 70% corn, 10% malted wheat, 10% honey malted wheat, 10% malted barley
Color: Light Straw
Price: $$$

Mike

Nose: Butter and caramel with malt.
Taste: Apples and caramel with some baking spices.
Finish: Citrus and spices.

Susan

Nose: Distinctly sweet malt, almost like a fruity Scotch.
Taste: Some chocolate notes, but mostly grains with a touch of spices.
Finish: Warm and lingering.

Notes

Currently contract distilled by New Riff in Newport, Kentucky. Rabbit Hole is building a distillery in downtown Louisville, scheduled to open in early 2018.

My Score

Notes: ____

____

____

Appearance (1-5) ____ Nose (1-25) ____
Taste (1-25) ____ Finish (1-25) ____
Complexity (1-10) ____ Overall Impression (1-10) ____
SCORE ____

Ranch Road Straight Bourbon Whiskey

Proof: 94
Age: 2 years
Type: Straight
Style: Small Batch
Mash Bill: Not available. Described as "rye-focused."
Color: Pale Straw
Price: $$

Mike

Nose: Vanilla and corn with a hint of ripe apple.
Taste: Vanilla, corn, apples and pears, with pepper and oak.
Finish: Long and dry with oak and pepper.

Susan

Nose: Ripe pears, honeysuckle, sweet spices, and vanilla.
Taste: Sweet mint, flowers, vanilla.
Finish: Sweet with a little heat. Pear lingers on the tongue.

Notes

Chankaska Creek Winery & Distillery, Kasota, Minnesota. Batch No. 5, Bottle No. 70. Longer aged bottles of this are now on the market.

My Score

Notes: ____________________________________

____

____

Appearance (1-5) ____ Nose (1-25) ____
Taste (1-25) ____ Finish (1-25) ____
Complexity (1-10) ____ Overall Impression (1-10) ____

SCORE ____

Rare Character Founder's Choice

Proof: 118.48
Age: 5 years, 2 months
Type: Straight
Style: Single Barrel
Mash Bill: 60% corn, 36% rye, 4% malted barley
Color: Light Amber
Price: $$$$

Mike

Nose: Vanilla, orange peel and lemon, some apricot, baking spice, and oak.
Taste: Vanilla, orange zest, and lemon with some raspberry and peach. Also, cinnamon, nutmeg, and oak. Water brings out the raspberry and a little cherry.
Finish: Long with oak and spice.

Susan

Nose: Peaches and citrus with some brown sugar, and very faint spice.
Taste: Fruit forward with caramel, a bit of vanilla, and baking spices.
Finish: Long and spicy.

Notes

Rare Character Whiskey sources from many states and bottles in Stanford, Kentucky. This one distilled in Indiana. Look limited releases including Old Man Bear and Raised By Wolves.

My Score

Notes: ____

____

____

Appearance (1-5) ____ Nose (1-25) ____
Taste (1-25) ____ Finish (1-25) ____
Complexity (1-10) ____ Overall Impression (1-10) ____
SCORE ____

RD One Double Finished in Oak and Maple

Proof:	99.9
Age:	NAS
Type:	Straight
Style:	Finished
Mash Bill:	70% corn, 21% rye, 9% malted barley
Color:	Dark Straw
Price:	$$$

Mike

Nose: Vanilla, maple syrup, oak.
Taste: Vanilla, maple syrup, cinnamon French toast, oakwood.
Finish: Oak and maple French toast.

Susan

Nose: Indian spices – curry, turmeric, cardamom. Some ripe apples and maple syrup.
Taste: Curry with golden raisins and caramel.
Finish: Maple and curry.

Notes

RD1 Spirits (Registered Distillery 1), Lexington, Kentucky.

My Score

Notes: ______________________________

____

____

Appearance (1-5) ____ Nose (1-25) ____
Taste (1-25) ____ Finish (1-25) ____
Complexity (1-10) ____ Overall Impression (1-10) ____

SCORE ____

RD One Finished with Brazilian Amburana Wood

Proof: 110
Age: NAS
Type: Straight
Style: Finished, Small Batch, Barrel Strength
Mash Bill: 70% corn, 21% rye, 9% malted barley
Color: Straw
Price: $$$

Mike

Nose: Vanilla, cinnamon red hots, oak.
Taste: Vanilla, cinnamon red hots candy, banana, oak.
Finish: Long with oak, cinnamon, and lingering banana.

Susan

Nose: Warm cinnamon toast topped with chopped baked apples.
Taste: Cinnamon sugar, apples, caramel. Breakfast pastry!
Finish: Sweet with cinnamon and brown sugar.

Notes

RD1 Spirits (Registered Distillery 1), Lexington, Kentucky.

My Score

Notes: ____

____

____

Appearance (1-5) ____ Nose (1-25) ____
Taste (1-25) ____ Finish (1-25) ____
Complexity (1-10) ____ Overall Impression (1-10) ____
SCORE ____

RD One Finished with French Oak

Proof: 101
Age: NAS
Type: Straight
Style: Finished, Small Batch
Mash Bill: 70% corn, 21% rye, 9% malted barley
Color: Dark Straw
Price: $$$

Mike

Nose: Caramel, milk chocolate, oak.
Taste: Caramel, dates, milk chocolate, cinnamon, oak.
Finish: Long with oak, cinnamon, and lingering chocolate.

Susan

Nose: Allspice, caramel, apple, cocoa.
Taste: Caramel apple, delicate spices, light oak.
Finish: Long with a chocolate note.

Notes

RD1 Spirits (Registered Distillery 1), Lexington, Kentucky.

My Score

Notes: ____________________________________

____

____

Appearance (1-5) ____ Nose (1-25) ____
Taste (1-25) ____ Finish (1-25) ____
Complexity (1-10) ____ Overall Impression (1-10) ____

SCORE ____

Rebellion

Proof: 90
Age: 8 years
Type: Straight
Style: Traditional
Mash Bill: Not available
Color: Light Straw
Price: $$

Mike

Nose: Corn, vanilla, and smoke with a hint of cherries.
Taste: Corn, vanilla, maple smoke. Hint of fruit.
Finish: Long and dry with oak and smoke.

Susan

Nose: Honey, caramel corn, woodsmoke.
Taste: Oranges, caramel, some roasted corn.
Finish: Dries to oak.

Notes

Distilled in Tennessee. Bottled by Market Street Spirits, Bardstown, Kentucky. This was bottle #10,106 of batch #1.

My Score

Notes: ______________________________

____

____

Appearance (1-5) ____ Nose (1-25) ____
Taste (1-25) ____ Finish (1-25) ____
Complexity (1-10) ____ Overall Impression (1-10) ____

SCORE ____

Redemption

Proof: 92
Age: 3 years
Type: Straight Bourbon
Style: Small Batch
Mash Bill: 60% corn, 30.2% rye, 1.8% barley
Color: Dark Straw
Price: $$

Mike

Nose: Corn and vanilla with a bit of honey. Very simple nose.
Taste: Corn and vanilla with some pepper spice. A bit thin.
Finish: Long and spicy with pepper spice and oak tannins.

Susan

Nose: Caramel and vanilla. A little sweet spice.
Taste: Vanilla, corn, some spice. Rather simple.
Finish: Long and warm. Peppery ending.

Notes

Rather surprising that the spice one expects from a high rye bourbon only comes to the fore in the finish. Distilled in Indiana and bottled in Bardstown, Kentucky by Strong Spirits.

My Score

Notes: ______________________________________

____

____

Appearance (1-5) ____ Nose (1-25) ____
Taste (1-25) ____ Finish (1-25) ____
Complexity (1-10) ____ Overall Impression (1-10) ____

SCORE ____

Remus Repeal Reserve

Proof:	94
Age:	NAS
Type:	Straight
Style:	Blend of straight bourbons
Mash Bill:	See notes
Color:	Straw
Price:	$$$

Mike

Nose: Apricots with vanilla and a little oak and spice.

Taste: Vanilla and apricots, cinnamon, nutmeg, and oak.

Finish: Long with oak and cinnamon and a little lingering apricot.

Susan

Nose: Vanilla sprinkled with herbs, figs and raisins, some tobacco leaf.

Taste: Vanilla and tobacco with sweet dates.

Finish: Long and peppery.

Notes

Blend of 50% 2005 bourbon (21% rye), 15% 2006 (36% rye), and 35% 2006 (21% rye). Made by MGP, Lawrenceburg, Indiana. Pair with a cigar.

My Score

Notes: ____________________________________

____

____

Appearance (1-5)	____	Nose (1-25)	____
Taste (1-25)	____	Finish (1-25)	____
Complexity (1-10)	____	Overall Impression (1-10)	____

SCORE ____

Robert's Reserve

Proof:	92
Age:	4 years
Type:	Straight
Style:	Pot Distilled, Small Batch, Four Grain
Mash Bill:	Not available
Color:	Light Amber
Price:	$$

Mike

Nose: Corn and caramel, with notes of malt, apricot, chocolate, and oak.

Taste: Corn, vanilla, chocolate, pepper spice, apricots, and oak.

Finish: Long with oak, pepper, and lingering chocolate.

Susan

Nose: Burnt caramel, root beer, dark fruit, and light spice.

Taste: Toffee, walnuts, dates and some apple peel with a sprinkling of black pepper.

Finish: Dries to warmth in the back of the throat.

Notes

Pauley Hollow Distillery, Forest Hills, Kentucky.

My Score

Notes: ______________________________

____

____

Appearance (1-5) ____ Nose (1-25) ____

Taste (1-25) ____ Finish (1-25) ____

Complexity (1-10) ____ Overall Impression (1-10) ____

SCORE ____

Rock Castle

Proof: 114.5
Age: 2 years
Type: Straight
Style: Cask Strength
Mash Bill: Not available
Color: Dark Straw
Price: $$$

Mike

Nose: Corn, a little caramel with a hint of oak.
Taste: Corn, vanilla, pepper spice, oak.
Finish: Long and spicy.

Susan

Nose: Caramel, vanilla, berries, and a touch of chocolate.
Taste: Milk chocolate up front followed by vanilla and red currants and a bit of cinnamon.
Finish: Ends in sweet oak.

Notes

Barrel House Distilling Co., Lexington, Kentucky. Another bourbon expression is Barrel House Select, 120 proof.

My Score

Notes: ______________________________

____

____

Appearance (1-5) ____ Nose (1-25) ____
Taste (1-25) ____ Finish (1-25) ____
Complexity (1-10) ____ Overall Impression (1-10) ____

SCORE ____

Rock Town Arkansas Bourbon Whiskey

Proof:	92
Age:	Under 4 years
Type:	Not Straight
Style:	Small Batch, Wheated
Mash Bill:	Not available
Color:	Very Light Amber
Price:	$$

Mike

Nose: Corn bread dough. Very yeasty with a hint of honey.
Taste: Corn bread and honey with a hint of oak.
Finish: Oaky and dry.

Susan

Nose: Cereal grain, dried corn. Touch of sugar.
Taste: Sweet oak. Some faint vanilla and peach.
Finish: Short and oaky.

Notes

From Rock Town Distillery, Little Rock, Arkansas. Not complex, but nicely made. Corn and wheat sourced from Arkansas.

My Score

Notes: ____

____

____

Appearance (1-5)	____	Nose (1-25)	____
Taste (1-25)	____	Finish (1-25)	____
Complexity (1-10)	____	Overall Impression (1-10)	____

SCORE ____

Rockfilter Distillery Fence Jumper

Proof: 94
Age: NAS
Type: Straight
Style: Two Grain
Mash Bill: Oaxacan green corn and cherrywood smoked winter rye
Color: Light Amber
Price: $$

Mike

Nose: Vanilla, dates, and cherries with some oak and baking spice.
Taste: Vanilla, baking spices, dates, and a little oak.
Finish: Medium long with some oak and spice.

Susan

Nose: Ripe corn, vanilla, cherries, cinnamon.
Taste: Green apples and black pepper with some light, smoked cherry.
Finish: Dries to oak and pepper.

Notes

Certified farm distillery. Grains are organically grown. Enzymes added in place of malted barley.

My Score

Notes: ______________________________

____

____

Appearance (1-5) ____ Nose (1-25) ____
Taste (1-25) ____ Finish (1-25) ____
Complexity (1-10) ____ Overall Impression (1-10) ____
SCORE ____

Rockfilter Distillery Giants of the Earth

Proof:	92
Age:	2 years, 5 months
Type:	Straight
Style:	Uses sorghum
Mash Bill:	Dwarf white sorghum, corn, winter rye
Color:	Straw
Price:	$$

Mike

Nose: Vanilla and apricots with a hint of oak.

Taste: Vanilla and pepper with a hint of apricots and oak.

Finish: Long, dry, and spicy with oak and pepper.

Susan

Nose: Apples and pears with some roasted born and faint molasses.

Taste: Caramel apple with vanilla and brown sugar.

Finish: Medium long and smooth.

Notes

Certified farm distillery. Grains are organically grown. Enzymes added in place of malted barley.

My Score

Notes: ______________________________

____

____

Appearance (1-5) ____ Nose (1-25) ____

Taste (1-25) ____ Finish (1-25) ____

Complexity (1-10) ____ Overall Impression (1-10) ____

SCORE ____

Rockfilter Distillery Rail Splitter

Proof: 94
Age: 2 years
Type: Straight
Style: Non-traditional flavoring grain
Mash Bill: 55% corn, 45% triticale
Color: Dark Straw
Price: $$

Mike

Nose: Vanilla and orange zest with a hint of oak.
Taste: Vanilla, orange zest, some baking spices, and oak.
Finish: Medium long with citrus and oak.

Susan

Nose: Vanilla and some light pepper with citrus notes and new leather.
Taste: Multigrain cereal, both sweet and spicy, with new leather, and dark fruit.
Finish: Dark fruit persists through the finish.

Notes

Rockfilter Distillery, Spring Grove, Minnesota. It's a Certified Farm Distillery.

My Score

Notes: ______________________________

____

____

Appearance (1-5) ____ Nose (1-25) ____
Taste (1-25) ____ Finish (1-25) ____
Complexity (1-10) ____ Overall Impression (1-10) ____

SCORE ____

Rockfilter Distillery Stone's Throw

Proof: 94
Age: 1 year, 8 months
Type: Not straight
Style: Uses oats
Mash Bill: Corn, oats, winter rye
Color: Straw
Price: $$

Mike

Nose: Corn, vanilla, and apricot with a hint of oak.
Taste: Vanilla, apricots, nutmeg, ginger, oak.
Finish: Long and spicy with a hint of oak.

Susan

Nose: Herbal with some sweet spices and stewed fruit.
Taste: Spiced oatmeal with a dusting of herbs.
Finish: Warm with set oak.

Notes

Organic grains. Enzymes added in place of malted barley.

My Score

Notes: ______________________________
____
____

Appearance (1-5) ____ Nose (1-25) ____
Taste (1-25) ____ Finish (1-25) ____
Complexity (1-10) ____ Overall Impression (1-10) ____

SCORE ____

Rod & Rifle

Proof: 80
Age: 4 years
Type: Straight
Style: Tennessee Bourbon
Mash Bill: 70% Corn, 25% Rye, 5% Malted Barley
Color: Very Light Straw
Price: $$

Mike

Nose: Very young. Corn with a hint of cherries.
Taste: Sweet corn and vanilla with a bit of pepper spice.
Finish: Slightly sweet with corn and dries out with some oak.

Susan

Nose: Caramel and corn. Some mustiness. Alcohol.
Taste: Sweet on the tongue. Very smooth. Vanilla predominates. Very simple.
Finish: Short, but smooth and warm. A little honey emerges at the very end.

Notes

Georgia Distilling Company/Prestige Imports. Expensive, but part of the proceeds go toward preserving hunting and fishing habitats.

My Score

Notes: ______________________________

____

____

Appearance (1-5) ____ Nose (1-25) ____
Taste (1-25) ____ Finish (1-25) ____
Complexity (1-10) ____ Overall Impression (1-10) ____
SCORE ____

Rough Rider

Proof: 88
Age: Minimum 2 years
Type: Straight
Style: Small Batch, Special Finish
Mash Bill: Not available
Color: Straw
Price: $$

Mike

Nose: Fruity with grapes and pine wood.
Taste: Sweet vanilla and grapes with some spice. Water takes the spice out.
Finish: Short and peppery. Water lessens the pepper and brings out more oak.

Susan

Nose: Caraway, rye, raisons, dates.
Taste: Caramel and cashews. Water reveals dates fruit and amplifies nuttiness.
Finish: Short and smooth. Bit of pepper at the end.

Notes

Long Island Spirits, Baiting Hollow, New York. Sourced from MGP, Lawrenceburg, Indiana. Second barreling in French oak casks that have rinsed with high proof brandy.

My Score

Notes: ____

____

____

Appearance (1-5) ____ Nose (1-25) ____
Taste (1-25) ____ Finish (1-25) ____
Complexity (1-10) ____ Overall Impression (1-10) ____

SCORE ____

Saint Paul Straight Bourbon

Proof: 84
Age: 3 years
Type: Straight
Style: Traditional
Mash Bill: Not available, but high rye
Color: Very Pale Straw
Price: $$

Mike

Nose: Corn, vanilla, hint of apricots.
Taste: Corn, vanilla, apricots and pepper with a hint of oak.
Finish: Long and dry with oak and pepper.

Susan

Nose: Candy corn, ripe apple, some fresh hay.
Taste: Vanilla and sweet corn, cinnamon, and some fruit.
Finish: Long and smooth.

Notes

Panther Distillery, Osakis, Minnesota. Also makes Ordeal Straight Bourbon, 4 years old and 80 proof.

My Score

Notes: ______________________________________

____

____

Appearance (1-5) ____ Nose (1-25) ____
Taste (1-25) ____ Finish (1-25) ____
Complexity (1-10) ____ Overall Impression (1-10) ____
SCORE ____

Six & Twenty 5 Grain

Proof: 90
Age: NAS
Type: Not Straight
Style: Small Batch
Mash Bill: See notes
Color: Dark Straw
Price: $$$

Mike

Nose: Sweet honey and peached. Water makes candied fruit come forward.
Taste: Honey and fruit. Some baking spices. Water dilutes the flavor.
Finish: Sweet and spicy with some oak.

Susan

Nose: Sweet. Candy corn and whiff of mint.
Taste: Light vanilla, whisper of orchard fruit. Some toasted marshmallow.
Finish: Mellow with a touch of black pepper at the end.

Notes

From Six & Twenty Distillery, Piedmont, South Carolina. Very enjoyable sip. Why not straight? Website says it is aged 2 years. Mash bill: Corn, soft red winter wheat, rye, barley, rice. Bottle #33 from a small batch of 8 barrels.

My Score

Notes: ____________________________________

____

____

Appearance (1-5) ____ Nose (1-25) ____
Taste (1-25) ____ Finish (1-25) ____
Complexity (1-10) ____ Overall Impression (1-10) ____
SCORE ____

Smooth Ambler Big Level Wheated Bourbon

Proof: 100
Age: 5+ years
Type: Straight
Style: Wheated, Non-Chill Filtered
Mash Bill: 71% corn, 21% wheat, 8% malted barley
Color: Straw
Price: $$$

Mike

Nose: Cereal grain, vanilla, with apples and hazelnuts.
Taste: Corn, vanilla, apple, and pepper spice.
Finish: Long and spicy.

Susan

Nose: Stoned wheat crackers, vanilla, pear.
Taste: Wheat crackers here, too. With some cinnamon and vanilla.
Finish: Starts sweet and dries to black pepper.

Notes

Smooth Ambler Spirits, Maxwelton, West Virginia, https://smoothambler.com. Add a little water to bring out more fruit.

My Score

Notes: ______________________________

____

____

Appearance (1-5) ____ Nose (1-25) ____
Taste (1-25) ____ Finish (1-25) ____
Complexity (1-10) ____ Overall Impression (1-10) ____
SCORE ____

Smooth Ambler Contradiction

Proof:	100
Age:	2 years
Type:	Mingling of Straight Bourbons
Style:	Traditional
Mash Bill:	Not available
Color:	Light Amber
Price:	$$$

Mike

Nose: Fruit and caramel. Ripe apples, baking spices and oak.
Taste: Apples, caramel. pepper spice and oak.
Finish: Long and dry with oak and pepper.

Susan

Nose: Caramel, vanilla, baking spices, nuts, apples.
Taste: Fruity, nutty caramel candy bar.
Finish: Smooth with peppery oak at the end.

Notes

Best 2-year-old ever! Smooth Ambler Spirits, Maxwelton, West Virginia. Bourbons produced both here and at MGP in Indiana were used. Majority share of this small distillery was purchased by liquor giant Pernod Richard in 2016.

My Score

Notes: ____

____

____

Appearance (1-5) ____ Nose (1-25) ____
Taste (1-25) ____ Finish (1-25) ____
Complexity (1-10) ____ Overall Impression (1-10) ____

SCORE ____

Smooth Ambler Old Scout

Proof: 99
Age: 6 years
Type: Straight
Style: Traditional
Mash Bill: Not available
Color: Dark Straw
Price: $$$

Mike

Nose: Fruit and caramel with a hint of tobacco.
Taste: Ripe apples and pear with vanilla and oak.
Finish: Starts sweet with some pepper and dries out nicely.

Susan

Nose: Sweet mint, anise, baking spice, vanilla.
Taste: Crème brulee, cinnamon, peaches and cherries.
Finish: Starts sweet with a dash of pepper at the end.

Notes

This was actually produced at MGP in Indiana. Smooth Ambler describes it as "fine curated whiskey scouted," hence the company name. But they are making and releasing more of their own whiskies as they have aged.

My Score

Notes: ____________________________________

____

____

Appearance (1-5) ____ Nose (1-25) ____
Taste (1-25) ____ Finish (1-25) ____
Complexity (1-10) ____ Overall Impression (1-10) ____

SCORE ____

Smooth Ambler Old Scout 10-Year-Old

Proof: 100
Age: 10 years
Type: Straight
Style: Small Batch
Mash Bill: Not available
Color: Dark Straw
Price: $$$

Mike

Nose: Caramel and molasses with tobacco, pecans and spice. Water brings out some chocolate.
Taste: Sweet caramel and apricots. A bit of oak and tobacco. Add water for fruit.
Finish: Medium long and oaky.

Susan

Nose: Vanilla, pitted fruit, toasted nuts, tobacco. Very nice.
Taste: Caramel, dates, ripe peach. Water adds some sweetness.
Finish: Medium long with pepper at the end.

Notes

Smooth Ambler, Maxwelton, West Virginia. Sourced from MGP, Lawrenceburg, Indiana. Small Batch No. 23, bottled 11/18/2014.

My Score

Notes: ______________________________

____

____

Appearance (1-5) ____ Nose (1-25) ____
Taste (1-25) ____ Finish (1-25) ____
Complexity (1-10) ____ Overall Impression (1-10) ____
SCORE ____

Smooth Ambler Old Scout - The Lion's Share Collection

Proof: 93
Age: 10 years
Type: Straight
Style: Single Barrel
Mash Bill: Not available
Color: Dark Straw
Price: $$$

Mike

Nose: Caramel, fruit and baking spice.
Taste: Caramel and milk chocolate with some berries and nutmeg.
Finish: Starts sweet and dries with some pepper and spices.

Susan

Nose: Rich candy corn and honey.
Taste: Sweet orchard fruit, nutmeg, sweet leather.
Finish: Very smooth and sweet. Touch of spice.

Notes

Distilled at MGP in Indiana. This was barrel #939. Excellent sip.

My Score

Notes: ______________________________

____

____

Appearance (1-5) ____ Nose (1-25) ____
Taste (1-25) ____ Finish (1-25) ____
Complexity (1-10) ____ Overall Impression (1-10) ____
SCORE ____

Smooth Ambler Old Scout Single Barrel

Proof:	112.2
Age:	11 years
Type:	Straight
Style:	Single Barrel, Barrel Strength, Extra Aged
Mash Bill:	Not available
Color:	Light Amber
Price:	$$$

Mike

Nose: Berries and pitted fruit. Vanilla with a hint of chocolate.

Taste: Fruit and caramel. Hint of baking spice and chocolate.

Finish: Caramel and chocolate leading into oak.

Susan

Nose: Caramel, sweet spices, dark cherries.

Taste: Toffee, meringue, cocoa, sweet spices.

Finish: Long and sweet. pleasantly warm.

Notes

Distilled at MGP in Indiana. Single barrel #811. The proof and age can vary.

My Score

Notes: ______________________________

____

____

Appearance (1-5) ____ Nose (1-25) ____

Taste (1-25) ____ Finish (1-25) ____

Complexity (1-10) ____ Overall Impression (1-10) ____

SCORE ____

Southern Star Paragon Bottled-in-Bond

Proof: 100
Age: 4 years
Type: Straight
Style: Wheated, Bonded
Mash Bill: 70% corn, 16% wheat, 14% malted barley
Color: Dark Straw
Price: $$$

Mike

Nose: Caramel, apricots, baking spice, and oak.
Taste: Caramel, apricots and peaches, white pepper, and oak.
Finish: Medium long with oak, pepper, and lingering caramel.

Susan

Nose: Toffee with some vanilla highlights, plums, and light baking spices.
Taste: Caramel joined by milk chocolate, plums, and a good dash od cinnamon.
Finish: Persistent fruit is accompanied by sweet oak.

Notes

Southern Distilling company, Statesville, North Carolina. Lovely complexity. Fine addition to the small, but growing, number of wheated bourbons.

My Score

Notes: ____________________________________

____

____

Appearance (1-5) ____ Nose (1-25) ____
Taste (1-25) ____ Finish (1-25) ____
Complexity (1-10) ____ Overall Impression (1-10) ____
SCORE ____

Southern Star Paragon Single Barrel

Proof: 119.6
Age: NAS
Type: Straight
Style: Wheated, Single Barrel, Cask Strength
Mash Bill: 70% corn, 16% wheat, 14% malted barley
Color: Dark Straw
Price: $$$

Mike

Nose: Caramel, apricots, baking spice, oak, and a hint of corn husk.
Taste: Corn husk, caramel, apricots, and oak.
Finish: Medium long with oak, pepper, and corn husk.

Susan

Nose: Citrus peel, apples, vanilla, caramel.
Taste: Caramel and fruit with some cinnamon and oak.
Finish: Fruit dries to oak and pepper.

Notes

Add water and both chocolate and fruit cocktail emerge. Really sweetens the whiskey.

My Score

Notes: ____________________________________

____

____

Appearance (1-5) ____ Nose (1-25) ____
Taste (1-25) ____ Finish (1-25) ____
Complexity (1-10) ____ Overall Impression (1-10) ____
SCORE ____

Southern Star Standard High Rye

Proof: 90
Age: NAS
Type: Straight
Style: Traditional
Mash Bill: 60% corn, 36% rye, 4% malted barley
Color: Dark Straw
Price: $$$

Mike

Nose: Vanilla and corn with a hint of pears and baking spice.
Taste: Vanilla and pears with some baking spice and oak.
Finish: Starts sweet but dries out with oak and spice.

Susan

Nose: Vanilla with some apple peel and rye spice.
Taste: Vanilla, hazelnuts, sweet oak, ginger, and pear.
Finish: Long, drying to peppery oak.

Notes

Southern Distilling Company, Statesville, North Carolina. A great profile and proof both the sip and mix.

My Score

Notes: ______________________________

____

____

Appearance (1-5) ____ Nose (1-25) ____
Taste (1-25) ____ Finish (1-25) ____
Complexity (1-10) ____ Overall Impression (1-10) ____
SCORE ____

Spring 44 Single Barrel

Proof: 100
Age: NAS
Type: Straight
Style: Single Barrel
Mash Bill: Not available
Color: Straw
Price: $$$$

Mike

Nose: Caramel and leather with some berry fruit.
Taste: Berries and apples with caramel and baking spice.
Finish: Long with balance of caramel and oak.

Susan

Nose: Corn and bread dough. Faint vanilla.
Taste: Vanilla, some pear and sweet spices.
Finish: Long and peppery.

Notes

Sourced from Kentucky. Bottled by Spring 44 Distillery, Loveland, Colorado. Proof adjusted with Colorado Rocky Mountain spring water. This was barrel #28.

My Score

Notes: ______________________________

____

____

Appearance (1-5) ____ Nose (1-25) ____
Taste (1-25) ____ Finish (1-25) ____
Complexity (1-10) ____ Overall Impression (1-10) ____

SCORE ____

St. Augustine Florida Double Cask Bourbon

Proof: 92.8
Age: NAS
Type: Not straight
Style: Wheated, Secondary Barrel
Mash Bill: 60% corn, 18% Florida wheat, 22% malted barley
Color: Dark Straw
Price: $$$

Mike

Nose: Very light. Hints of vanilla and chocolate.
Taste: Caramel, pepper, and a hint of fruit.
Finish: Short and dry with pepper and oak.

Susan

Nose: Sweet corn. Light caramel.
Taste: Oak and caramel corn with some honey drops.
Finish: Very warm and dry.

Notes

Aged in new and used oak barrels. First release from this distillery. Other expressions include Florida Straight Bourbon and Port Finished bourbon. St. Augustine Distillery, Florida, https://www.staugustinedistillery.com.

My Score

Notes: ____

____

____

Appearance (1-5) ____ Nose (1-25) ____
Taste (1-25) ____ Finish (1-25) ____
Complexity (1-10) ____ Overall Impression (1-10) ____

SCORE ____

St. Augustine Florida Straight Bourbon

Proof: 88
Age: 3 years
Type: Straight
Style: Wheated
Mash Bill: 60% Florida corn, 18% wheat, 22% malted barley
Color: Amber
Price: $$

Mike

Nose: Corn, vanilla, hint of fruit and oak.
Taste: Corn, vanilla, pears, cinnamon, and oak.
Finish: Medium long with oak and cinnamon.

Susan

Nose: Roasted corn and vanilla with some light apple and faint baking spices.
Taste: Matches the nose with the addition of light oak tannins.
Finish: Dries to light oak.

Notes

St. Augustine Distillery, Florida. There's a port wine finished expression, too.

My Score

Notes: ____

____

____

Appearance (1-5) ____ Nose (1-25) ____
Taste (1-25) ____ Finish (1-25) ____
Complexity (1-10) ____ Overall Impression (1-10) ____

SCORE ____

Starlight Carl T. Huber's Bourbon

Proof: 95
Age: 4 years
Type: Straight
Style: Small Batch, Pot Distilled
Mash Bill: 58% corn, 27% rye, 15% malted barley
Color: Light Straw
Price: $$$

Mike

Nose: Corn, vanilla, apricots, berries, with some oak wood.
Taste: Corn, vanilla, berries, pepper, and oak.
Finish: Long and peppery with a little oak.

Susan

Nose: Caramel and butterscotch with a little banana and sweet mint.
Taste: Caramel, crème brulee, berries, and baking spices.
Finish: Tingly with pepper drying to oak.

Notes

Huber Winery Starlight Distillery, Borden, Indiana. Look for other expressions finished in ex-wine and ex-spirits barrels. All well-balanced.

My Score

Notes: ____

____

____

Appearance (1-5) ____ Nose (1-25) ____
Taste (1-25) ____ Finish (1-25) ____
Complexity (1-10) ____ Overall Impression (1-10) ____

SCORE ____

Starlight Carl T. Huber's Double Oaked

Proof: 106
Age: 4 ½ years
Type: Straight
Style: Secondary Barrel
Mash Bill: 58% corn, 27% rye, 15% malted barley
Color: Dark Straw
Price: $$$

Mike

Nose: Sorghum, plums and dates, baking spices and oak.
Taste: Caramel, plums, dates, cinnamon, nutmeg, and oak.
Finish: Long with oak and baking spice with a hint of sorghum.

Susan

Nose: Warm sorghum over cornbread and a little baking spice.
Taste: Burnt vanilla and sweet spices with dark fruit.
Finish: Lots of pepper but not hot.

Notes

Water softens it and releases more fruit. Additional aging in a toasted barrel for 6 months.

My Score

Notes: ____

____

____

Appearance (1-5) ____ Nose (1-25) ____
Taste (1-25) ____ Finish (1-25) ____
Complexity (1-10) ____ Overall Impression (1-10) ____

SCORE ____

Stave & Barrel

Proof: 118.8
Age: 6 months
Type: Not Straight
Style: Single Barrel, Barrel Proof
Mash Bill: 21% rye
Color: Straw
Price: $$$

Mike

Nose: Corn, marshmallows, vanilla.
Taste: Corn and vanilla with sweet spices.
Finish: Medium long, spicy.

Susan

Nose: Bananas, vanilla, some nutmeg.
Taste: Banana pudding with nutmeg.
Finish: Warm, drying to oak.

Notes

Distilled in Indiana. Bottled by Glenns Creek Distillery, Frankfort, Kentucky. Barrel #2, bottle #128.

My Score

Notes: ______________________________

____

____

Appearance (1-5) ____ Nose (1-25) ____
Taste (1-25) ____ Finish (1-25) ____
Complexity (1-10) ____ Overall Impression (1-10) ____
SCORE ____

The Steward's Solera Bourbon

Proof: 100
Age: See notes
Type: Straight
Style: Small Batch
Mash Bill: See notes
Color: Light Amber
Price: $$$

Mike

Nose: Vanilla and caramel. Apricot and light oak.
Taste: Sweet caramel and apricot with a lot of oak tannins to the point of bitterness.
Finish: Long – oaky – bitter.

Susan

Nose: Very nutty. Toasted almonds with some vanilla notes.
Taste: Rich mouthful. Palate matches nose with nuts and vanilla.
Finish: Long and warm. Pepper and oak on the end.

Notes

From The Steward's Solera, Aurora, Colorado. Concept is to create a "meritage" of bourbon. Eight bourbons from four states are mingled. At least some of this whiskey comes from MGP in Indiana (put in barrels in 2000). Other barrels from Kentucky (2000, 2001, 2004) and Tennessee (2007) are added to new craft distillate from this Colorado distillery. Mash bills between 60 and 75% corn, 20% rye, 5% malted barley.

My Score

Notes: ____

____

____

Appearance (1-5) ____ Nose (1-25) ____
Taste (1-25) ____ Finish (1-25) ____
Complexity (1-10) ____ Overall Impression (1-10) ____

SCORE ____

Still & Oak Straight Bourbon Whiskey

Proof: 86
Age: NAS
Type: Straight
Style: Non-Chill Filtered
Mash Bill: 67% corn, 11% rye, 22% malted barley
Color: Very, Very Pale Straw
Price: $$

Mike

Nose: Very light. Corn bread dough and vanilla.
Taste: Corn and vanilla with hints of pepper and oak.
Finish: Short with only a little oak.

Susan

Nose: Buttered corn, some apple, some faint spice.
Taste: Caramel and baking spices with some brown sugar and ripe apple.
Finish: Short and peppery.

Notes

Great lakes distillery, Milwaukee, Wisconsin. All grains used grown in Wisconsin.

My Score

Notes: ______________________________
____
____

Appearance (1-5) ____ Nose (1-25) ____
Taste (1-25) ____ Finish (1-25) ____
Complexity (1-10) ____ Overall Impression (1-10) ____
SCORE ____

Still Austin Cask Strength

Proof: 118
Age: 2 years
Type: Straight
Style: Cask Strength
Mash Bill: 70% white corn, 25% Elbon rye, 5% Wildfire malted barley
Color: Dark Straw
Price: $$$

Mike

Nose: Corn, vanilla, floral note, oak.
Taste: Corn, citrus, vanilla, baking spice, and oak.
Finish: Long with oak and cinnamon.

Susan

Nose: Caramel and honey, some sweet herbs and dark cherry.
Taste: Lavender honey, lemon, cinnamon, and brown sugar.
Finish: Dries quickly to oak and some leather.

Notes

Still Austin Whiskey Company, Austin, Texas. All grains Texas-sourced.

My Score

Notes: ______________________________

____

____

Appearance (1-5) ____ Nose (1-25) ____
Taste (1-25) ____ Finish (1-25) ____
Complexity (1-10) ____ Overall Impression (1-10) ____

SCORE ____

Stillwrights Bourbon

Proof: 90
Age: 2 years
Type: Straight
Style: Small Batch, Craft
Mash Bill: Not available
Color: Dark Straw
Price: $$

Mike

Nose: Vanilla and corn bread. Very light caramel.
Taste: Young – vanilla and cornbread dough and a hint of pepper spice.
Finish: Long with a bit of oak and pepper.

Susan

Nose: Corn, fusel oils, hayloft, faint with a hint of spice and oak.
Taste: Toffee and a little mint, hazelnuts.
Finish: Long with lots of black pepper at the end.

Notes

Overall quite young. Flat Rock Spirits of Fairborn, Ohio (near Dayton) is the distiller. They like to use locally sourced ingredients.

My Score

Notes: ______________________________

____

____

Appearance (1-5) ____ Nose (1-25) ____
Taste (1-25) ____ Finish (1-25) ____
Complexity (1-10) ____ Overall Impression (1-10) ____

SCORE ____

Stone Hammer Distiller's Reserve

Proof: 90
Age: NAS
Type: Straight
Style: Traditional
Mash Bill: Not available
Color: Light Straw
Price: $

Mike

Nose: Very light. Corn, vanilla, hint of oak.
Taste: Corn, vanilla, ripe apples, oak, and pepper.
Finish: Very spicy with lots of black pepper and a hint of oak.

Susan

Nose: Caramelized bananas, light brown sugar, baking spice.
Taste: Fruity caramel nut candy bar with a dash of cinnamon and allspice.
Finish: Long and warm. No bitterness.

Notes

Clear Springs Distillery, Louisville, Kentucky. This seems to be a label of Sazerac. It's a Kroger house brand. Quite good for its price.

My Score

Notes: ____

____

____

Appearance (1-5) ____ Nose (1-25) ____
Taste (1-25) ____ Finish (1-25) ____
Complexity (1-10) ____ Overall Impression (1-10) ____

SCORE ____

Straight Edge

Proof: 84
Age: NAS
Type: Not Straight
Style: Special Finish
Mash Bill: Not available
Color: Dark Straw
Price: $$$

Mike

Nose: Vanilla and honey with a hint of banana.
Taste: Light, not complex. Corn and vanilla. A hint of oak.
Finish: Medium long and dry with oak.

Susan

Nose: Corn, bread dough, mint.
Taste: Sweet vanilla and caramel corn.
Finish: Long and warm, verging on hot.

Notes

"Finished and bottled" by Splinter Group, Napa, California. Blend of 5, 7, and 8 year old Kentucky and Tennessee bourbons. Finished in cabernet sauvignon barrels.

My Score

Notes: ______________________________

____

____

Appearance (1-5) ____ Nose (1-25) ____
Taste (1-25) ____ Finish (1-25) ____
Complexity (1-10) ____ Overall Impression (1-10) ____
SCORE ____

Tatoosh Bourbon

Proof: 80
Age: 3 years
Type: Not Straight
Style: Small Batch
Mash Bill: Not available
Color: Very Light Straw
Price: $$

Mike

Nose: Maple smoke and vanilla.
Taste: Very light. Maple wood and vanilla with a hint of Flintstone chewable vitamins.
Finish: Quick and smoky.

Susan

Nose: Corn and light maple syrup.
Taste: Honeyed oak, sweet mint, hard candy.
Finish: Short and warm.

Notes

Finished and bottled by Tatoosh Distillery, Seattle, Washington. No indication of source on the label.

My Score

Notes: ____________________________________

____

____

Appearance (1-5) ____ Nose (1-25) ____
Taste (1-25) ____ Finish (1-25) ____
Complexity (1-10) ____ Overall Impression (1-10) ____

SCORE ____

Temperance Trader

Proof: 98.2
Age: NAS
Type: Straight
Style: Small Batch
Mash Bill: Not available
Color: Light Straw
Price: $$

Mike

Nose: Lots of corn. Hints of vanilla and oak.
Taste: Corn and vanilla with some pepper.
Finish: Lots of pepper and some oak.

Susan

Nose: Pecans, light oak, some sweet mint.
Taste: Nutty, with some vanilla. Very smooth.
Finish: Sweet with sweet oak and corn at the end.

Notes

Distilled in Tennessee. Bottled by Bull Run Distilling Co., Portland, Oregon. Batch #017 and bottle #04437. Enjoyable light sip. But even at 99.02 proof it would disappear in a cocktail.

My Score

Notes: ____

____

____

Appearance (1-5) ____ Nose (1-25) ____
Taste (1-25) ____ Finish (1-25) ____
Complexity (1-10) ____ Overall Impression (1-10) ____

SCORE ____

Temptation

Proof: 82
Age: 3 years
Type: Straight Bourbon
Style: Small Batch
Mash Bill: 75% corn, 20% rye, 5% barley
Color: Straw
Price: $$

Mike

Nose: Very light with corn and vanilla with a hint of spice and oak.
Taste: Sweet with sweet spices and corn. Just a hint of oak. A bit thin in the mouth.
Finish: Long and peppery turning dry with oak tannins.

Susan

Nose: Light nose. faint caramel, vanilla and spices.
Taste: Corn and pepper, some sweet spice. Not complex.
Finish: Hot and peppery. Dries quickly.

Notes

Distilled in Indiana and bottled in Bardstown, Kentucky by Bardstown Barrel Selection.

My Score

Notes: ______________________________
____
____

Appearance (1-5) ____ Nose (1-25) ____
Taste (1-25) ____ Finish (1-25) ____
Complexity (1-10) ____ Overall Impression (1-10) ____

SCORE ____

Thirteenth Colony Bourbon

Proof: 95
Age: NAS
Type: Not Straight
Style: Small Batch, Craft
Mash Bill: Not available
Color: Light Straw
Price: $$

Mike

Nose: Sweet corn and vanilla. A little honey, sweet spice.
Taste: A bit thin on the mouth. More pepper spice, then vanilla. Very astringent.
Finish: Very astringent and peppery finish. Not complex. Not good.

Susan

Nose: Very light. Some vanilla, corn, and touch of honey.
Taste: Corn and some pepper. Very simple.
Finish: Long, painfully hot, peppery.

Notes

Thirteenth Colony Distilleries, Americus, Georgia. Batch No. 2014. Bottle No. 3149.

My Score

Notes: ______________________________

____

____

Appearance (1-5) ____ Nose (1-25) ____
Taste (1-25) ____ Finish (1-25) ____
Complexity (1-10) ____ Overall Impression (1-10) ____

SCORE ____

Three Chord Twelve Bar Reserve

Proof: 107
Age: 12 years
Type: Straight
Style: Cask Strength, Small Batch
Mash Bill: 84% corn, 8% rye, 8% malted barley
Color: Straw
Price: $$$

Mike

Nose: Caramel and prunes with a hint of honey, baking spices, and oak in the background.
Taste: Caramel, prunes, dates, allspice, cardamom, some oak.
Finish: Long with oak, spice, and a bit of dark chocolate.

Susan

Nose: Very nutty with some toffee and vanilla icing.
Taste: Snickers candy bar. Little spice, dark fruit.
Finish: Warm with some pepper.

Notes

Steel Bending Distilling Co., Chelsea, Michigan. The distillery also has a blend of straight bourbons expression.

My Score

Notes: ____

____

____

Appearance (1-5) ____ Nose (1-25) ____
Taste (1-25) ____ Finish (1-25) ____
Complexity (1-10) ____ Overall Impression (1-10) ____
SCORE ____

Tom's Foolery Barrel 11

Proof: 112
Age: ~ 3 years
Type: Straight
Style: Single Barrel, Craft
Mash Bill: 57% corn, 7% rye, 36% vienna malt
Color: Dark Straw
Price: $$$

Mike

Nose: Corn, vanilla, and apricots.
Taste: Corn and anise with vanilla and spice. Very sweet, with nice mouthfeel. No burn at 112 proof!!!
Finish: Short and sweet.

Susan

Nose: Lightly toasted oak and touch of leather.
Taste: Sweeter on palate with ripe fruit, almonds and honey notes.
Finish: Short and sweet with mild pepper spice. Soft, oaky ending.

Notes

Tom's Foolery Distillery, Chagin Falls, Ohio. (Near Cleveland.) The owners bought some used equipment from the decommissioned Michter's Distiller in Schaeffertown, Pennsylvania.

My Score

Notes: ____________________

____

____

Appearance (1-5) ____ Nose (1-25) ____
Taste (1-25) ____ Finish (1-25) ____
Complexity (1-10) ____ Overall Impression (1-10) ____

SCORE ____

Tom's Foolery Barrel 19

Proof:	120
Age:	~ 3 ½ years
Type:	Straight
Style:	Single Barrel, Craft
Mash Bill:	73% corn, 13% rye, 12% malted barley
Color:	Dark Straw
Price:	$$$

Mike

Nose: Corn and honey with a hint of only a little oak. Water complex, but pleasant. Softens the nose and brings out candy corn.

Taste: Corn and yeasty bread dough. Vanilla and pepper spice. Surprisingly not hot at barrel proof. Water brings out more vanilla.

Finish: Sweet and long with honey and pecans.

Susan

Nose: Corn! And a little pitted fruit. Not vanilla.

Taste: Candies nuts, new saddle leather, peppery spice. Water amplifies the leather.

Finish: Very long. Add a little water to tame the proof heat.

Notes

Single Barrel bottling from Tom's Foolery Distillery near Cleveland. Nice layers for a young bourbon.

My Score

Notes: ______________________________

____

____

Appearance (1-5)	____	Nose (1-25)	____
Taste (1-25)	____	Finish (1-25)	____
Complexity (1-10)	____	Overall Impression (1-10)	____

SCORE ____

Tom's Foolery Barrel 118

Proof:	108
Age:	3 years
Type:	Straight
Style:	Single Barrel, Craft
Mash Bill:	52% white corn, 10% rye, 32% barley*
Color:	Light Straw
Price:	$$$

Mike

Nose: Very light. Vanilla and corn with some peaches.

Taste: Corn and spice – anise and vanilla. peach or apricot and just a hint of oak.

Finish: Medium length. Sweet and spicy.

Susan

Nose: Soft oak, like woodshop shavings.

Taste: Quite sweet with vanilla, sweet Some ripe spices and a hint dash of candied nuts.

Finish: Medium long and smooth. No burn. Perhaps its outstanding feature.

Notes

Each single barrel from Tom's Foolery is different. This one employs a trio of barleys* – 29% barley malt, 6% crystal malt, 3% roasted barley.

My Score

Notes: ______________________________

____

____

Appearance (1-5)	____	Nose (1-25)	____
Taste (1-25)	____	Finish (1-25)	____
Complexity (1-10)	____	Overall Impression (1-10)	____

SCORE ____

Tom's Foolery Small Batch 2015.1

Proof:	90
Age:	2 ½ years
Type:	Straight
Style:	Small Batch
Mash Bill:	Not available
Color:	Light Straw
Price:	$$$

Mike

Nose: Very young. Corn and vanilla. Hint of peaches.

Taste: Very nice. Corn and peaches with some cinnamon.

Finish: Slightly oaky with spices and vanilla.

Susan

Nose: Corn predominates. Yeasty.

Taste: Sweet corn at first. Then caramel and vanilla.

Finish: Short, sweet, warm, and smooth.

Notes

From Tom's Foolery Distillery, Cleveland, Ohio. Not complex, but a very pleasant sip. Bottle #1185.

My Score

Notes: ______________________________

____

____

Appearance (1-5) ____ Nose (1-25) ____

Taste (1-25) ____ Finish (1-25) ____

Complexity (1-10) ____ Overall Impression (1-10) ____

SCORE ____

Trail's End Kentucky Straight Bourbon

Proof: 90
Age: 8 years
Type: Straight
Style: Secondary Barrel
Mash Bill: Not available
Color: Dark Straw
Price: $$$

Mike

Nose: Vanilla and leather with a hint of oak.
Taste: A bit thin, corn husks, vanilla, a bit of pitted fruit.
Finish: Long, dry with oak and pepper.

Susan

Nose: Very caramel, with book leather, baking spice, and dark fruit.
Taste: Caramel, dark cherries, cinnamon, nutmeg, ginger. Rich mouthfeel.
Finish: Long and complex.

Notes

Hood River Distillers, Inc., Hood River, Oregon. Sourced Kentucky bourbon finished in Oregon white oak (*Quercus garryana*) barrels.

My Score

Notes: ____________________

____

____

Appearance (1-5) ____ Nose (1-25) ____
Taste (1-25) ____ Finish (1-25) ____
Complexity (1-10) ____ Overall Impression (1-10) ____
SCORE ____

Traverse City Bourbon

Proof: 86
Age: 4 years
Type: Straight
Style: Small Batch
Mash Bill: Not available
Color: Straw
Price: $$$

Mike

Nose: Light with corn, vanilla and a hint of oak.
Taste: Sweet vanilla with corn, a hint of oak and fruit, and a little pepper.
Finish: Long and dry with oak and pepper.

Susan

Nose: Corn, vanilla, a little ripe pear.
Taste: Vanilla forward with some fruit and nuts.
Finish: Dried to peppery oak.

Notes

Traverse City Whiskey Co., Michigan. Nicely balanced.

My Score

Notes: ____________________________________

____

____

Appearance (1-5) ____ Nose (1-25) ____
Taste (1-25) ____ Finish (1-25) ____
Complexity (1-10) ____ Overall Impression (1-10) ____
SCORE ____

Underground Black Cherry Wood Finished

Proof: 94
Age: NAS
Type: Straight
Style: Finished
Mash Bill: Not available
Color: Light Amber
Price: $$$

Mike

Nose: Vanilla and oak with a little sweet honey.
Taste: Very young. Corn, vanilla, honey, wood.
Finish: Long and astringent with wood.

Susan

Nose: Cherry cough syrup.
Taste: Dark cherries, some vanilla, sweet oak.
Finish: Sweet with some light pepper.

Notes

Cleveland Whiskey, Ohio.

My Score

Notes: ______________________________

____

____

Appearance (1-5) ____ Nose (1-25) ____
Taste (1-25) ____ Finish (1-25) ____
Complexity (1-10) ____ Overall Impression (1-10) ____

SCORE ____

Waitsburg Bourbon Whiskey

Proof: 80
Age: NAS
Type: Straight
Style: Small Batch, Craft
Mash Bill: Not available
Color: Light Straw
Price: $$

Mike

Nose: Vanilla and sweet corn. Very light with just a hint of oak.
Taste: Sweet spice and vanilla. Not complex. Some corn and oak.
Finish: Moderately long. A bit spicy with oak tannins.

Susan

Nose: Minty, light, with a whiff of spice.
Taste: Light oak with some spice and a little vanilla.
Finish: Short and quite peppery for 80 proof.

Notes

From the Oola Distillery in Seattle, Washington. Craft distiller makes other spirits, as well, including gin, vodka, and infused vodkas.

My Score

Notes: ____

____

____

Appearance (1-5) ____ Nose (1-25) ____
Taste (1-25) ____ Finish (1-25) ____
Complexity (1-10) ____ Overall Impression (1-10) ____
SCORE ____

Watershed

Proof:	94
Age:	NAS
Type:	Not Straight
Style:	Small Batch
Mash Bill:	Corn, wheat, rye, spelt
Color:	Straw
Price:	$$

Mike

Nose: Candied corn with a hint of spice.
Taste: Corn and vanilla with some pepper.
Finish: Long and dry with oak. A bit astringent.

Susan

Nose: Candied apples and some burnt vanilla.
Taste: Buttered popcorn, green apples.
Finish: Short. Not complex.

Notes

From Watershed Distillery, Columbus, Ohio. Small batch #013.

My Score

Notes: ____

____

____

Appearance (1-5) ____ Nose (1-25) ____
Taste (1-25) ____ Finish (1-25) ____
Complexity (1-10) ____ Overall Impression (1-10) ____
SCORE ____

Watershed Bottled-in-Bond

Proof: 100
Age: 4 years
Type: Straight
Style: Bonded
Mash Bill: Not available
Color: Pale Straw
Price: $$

Mike

Nose: Very fruity. Berries and apricots with vanilla and oak.
Taste: More citrus than suggested by the nose. Lemon zest and apricots with a little cinnamon, nutmeg, and sweet oak.
Finish: Long with oak and spice and lingering lemon zest.

Susan

Nose: Fresh green apples, spicy herbs, some light vanilla.
Taste: Savory. Salted caramel, lemon drizzled apple and some pepper.
Finish: Long and with peppery oak.

Notes

Watershed Distillery, Columbus, Ohio. Nicely balanced sip.

My Score

Notes: ____

____

____

Appearance (1-5) ____ Nose (1-25) ____
Taste (1-25) ____ Finish (1-25) ____
Complexity (1-10) ____ Overall Impression (1-10) ____
SCORE ____

Watershed Bourbon - Apple Brandy Finished

Proof: 90
Age: 4 years
Type: Straight
Style: Finished
Mash Bill: Not available
Color: Pale Straw
Price: $$

Mike

Nose: Apples and vanilla, hints of baking spice and oak.
Taste: Apples – both ripe and tart – with vanilla, allspice, and oak.
Finish: Medium long with oak, spice, and lingering tart apple.

Susan

Nose: Sweet grain, apples, berries, and a little ginger.
Taste: Fruit forward with berries, cherry preserves, apple, lemon peel, and a dusting of baking spice.
Finish: Dries to sweet oak.

Notes

Finished in apple brandy barrels, which is another spirit made by Watershed.

My Score

Notes: ______________________________

____

____

Appearance (1-5) ___ Nose (1-25) ___
Taste (1-25) ___ Finish (1-25) ___
Complexity (1-10) ___ Overall Impression (1-10) ___

SCORE ___

Wathen's

Proof: 94
Age: NAS
Type: Straight
Style: Single Barrel
Mash Bill: Not available
Color: Straw
Price: $$$

Mike

Nose: Orange blossom, honey, corn and oak. Not complex, but very nice.
Taste: Sweet caramel, oak and a bit of citrus fruit. Peppery spices and fine tobacco.
Finish: Long and dry. Spicy with oak and pepper.

Susan

Nose: Fruity, with notes of banana and paw paw. Little bit of underlying sweet spice and a dash of cocoa.
Taste: Cinnamon and sweet caramel with peppery notes.
Finish: Sweet at the beginning, which gradually dries to pepper and oak.

Notes

A very pleasant single barrel to sip neat or with a chip of ice. Goes well with a cigar or steak. Distilled and aged in Kentucky and bottled by Charles Medley Distillery, San Jose, California.

My Score

Notes: ______________________________

____

____

Appearance (1-5) ____ Nose (1-25) ____
Taste (1-25) ____ Finish (1-25) ____
Complexity (1-10) ____ Overall Impression (1-10) ____
SCORE ____

West of Kentucky No. 2

Proof: 95
Age: 1 year
Type: Unfiltered
Style: Small Batch, Wheated
Mash Bill: Not available
Color: Straw
Price: $$$

Mike

Nose: Very light. Bread dough and vanilla.
Taste: Corn and citrus with some vanilla and spice.
Finish: Dry oak and spice.

Susan

Nose: Cellulose. Some caramel and vanilla.
Taste: Corn, oak, vanilla.
Finish: Short and peppery.

Notes

Made by Sonoma County Distilling Co., Rohnert Park, California. Bottle #1243 from batch #3.

My Score

Notes: ______________________________

____

____

Appearance (1-5) ____ Nose (1-25) ____
Taste (1-25) ____ Finish (1-25) ____
Complexity (1-10) ____ Overall Impression (1-10) ____

SCORE ____

West of Kentucky No. 2 Cask Strength

Proof: 107
Age: 1 year
Type: Not Straight
Style: Wheated, Small Batch, Barrel Strength
Mash Bill: Not available
Color: Straw
Price: $$$

Mike

Nose: Bread dough and vanilla. Hint of oak.
Taste: Corn and citrus. Some pecans and oak.
Finish: Dry with oak and spice.

Susan

Nose: Corn husk. Nut shells.
Taste: Oak and roasted corn. Some dates.
Finish: Very oaky.

Notes

Made by Sonoma County Distilling Co., Rohnert Park, California. Bottle #248 from batch #3.

My Score

Notes: ____

____

____

Appearance (1-5) ____ Nose (1-25) ____
Taste (1-25) ____ Finish (1-25) ____
Complexity (1-10) ____ Overall Impression (1-10) ____
SCORE ____

Western Reserve

Proof: 86
Age: 4 years
Type: Straight
Style: Blend of straight bourbons
Mash Bill: Not available
Color: Light Amber
Price: $$$

Mike

Nose: Vanilla, citrus, a little baking spice, oak.
Taste: Vanilla and orange peel with some cinnamon and oak.
Finish: Medium long with oak and cinnamon.

Susan

Nose: Top notes of brown sugar and caramel with light apple.
Taste: Almost malty, with caramel, cinnamon, and apple.
Finish: Ends in a pop of black pepper.

Notes

Western Reserve Distillers, Cleveland, Ohio. Some 14-year-old in the blend. One of several expressions produced.

My Score

Notes: ______________________________

____

____

Appearance (1-5) ____ Nose (1-25) ____
Taste (1-25) ____ Finish (1-25) ____
Complexity (1-10) ____ Overall Impression (1-10) ____
SCORE ____

W.H. McBrayer

Proof: 103.6
Age: NAS
Type: Straight
Style: Small Batch, Cask Strength, Non-Chill Filtered
Mash Bill: 88.4% Bloody Butcher corn, 5.8% heritage rye, 5.8% malted barley
Color: Light Amber
Price: $$$

Mike

Nose: Caramel and apricots with a little oak and leather.
Taste: Caramel, apricots, lots of cinnamon and pepper, with a little oak.
Finish: Long and spicy with oak, baking spices.

Susan

Nose: Brown sugar, peeled peaches, dusting of cinnamon.
Taste: Chew caramel. Green apples, a little mint, pepper.
Finish: Long and warm.

Notes

McBrayer Legacy Spirits, Danville, Kentucky. Contract distilled at Wilderness Trail. (See Old McBrayer, too.)

My Score

Notes: ______________________________
____
____

Appearance (1-5) ____ Nose (1-25) ____
Taste (1-25) ____ Finish (1-25) ____
Complexity (1-10) ____ Overall Impression (1-10) ____
SCORE ____

W.H. McBrayer Barrel Strength

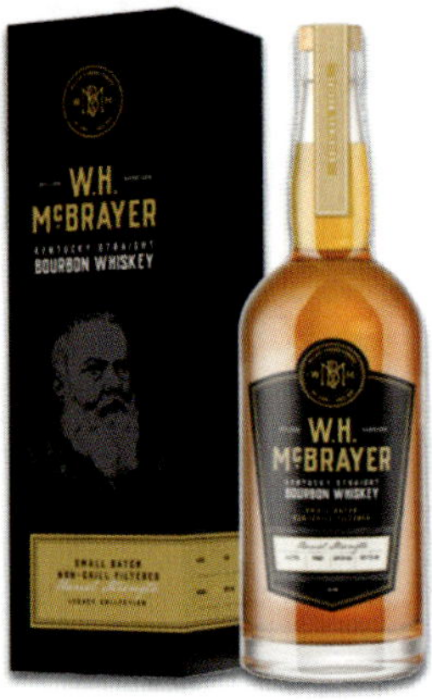

Proof: 103.5
Age: Blend of 4, 5, and 6 years
Type: Straight
Style: Barrel Strength, Small Batch, Non-Chill Filtered
Mash Bill: 88.4% corn, 5.8% rye, 5.8% malted barley
Color: Straw
Price: $$$$

Mike

Nose: Vanilla, marshmallows, dried fruit, baking spices, oak.
Taste: Vanilla, raisins, plums, cinnamon, nutmeg, oak.
Finish: Long with oak and cinnamon.

Susan

Nose: Vanilla, orchard fruit, nutmeg.
Taste: Vanilla, dark fruit, cinnamon, nutmeg, oak.
Finish: Long, warm, and peppery.

Notes

Complex and balanced. McBrayer Legacy Spirits, Bardstown, Kentucky.

My Score

Notes: ______________________________

____

____

Appearance (1-5) ____ Nose (1-25) ____
Taste (1-25) ____ Finish (1-25) ____
Complexity (1-10) ____ Overall Impression (1-10) ____

SCORE ____

W.H. McBrayer Cedar Brook Wheated Bourbon

Proof: 100
Age: NAS
Type: Straight
Style: Wheated
Mash Bill: 70% corn, 21% wheat, 9% malted barley
Color: Straw
Price: $$$

Mike

Nose: Vanilla, apples, hazelnuts, oak.
Taste: Vanilla, apples, hazelnuts, white pepper, oak.
Finish: Short with oak and pepper.

Susan

Nose: Apple, pears, vanilla, cashews, and some light cinnamon sugar.
Taste: Vanilla, range peel, cinnamon streusel, a little milk chocolate.
Finish: Nicely balanced between sweet elements and light oak.

Notes

McBrayer Legacy Spirits, Bardstown, Kentucky. William Harrison McBrayer of Lawrenceburg was a 19th century judge, farmer, distiller, senator, and philanthropist.

My Score

Notes: ____________________________________

____

____

Appearance (1-5) ____ Nose (1-25) ____
Taste (1-25) ____ Finish (1-25) ____
Complexity (1-10) ____ Overall Impression (1-10) ____
SCORE ____

The Wheater

Proof: 90
Age: 2 years
Type: Straight
Style: Wheated, Blend
Mash Bill: Not available
Color: Very Pale Straw
Price: $$$

Mike

Nose: Corn and bread dough with some vanilla and fruit.
Taste: Sweet corn, honey, vanilla, pears, apples, hints of pepper and oak.
Finish: Medium long with some pepper and oak.

Susan

Nose: Cornbread dough, baked apple.
Taste: Vanilla and wheat thins with some cinnamon.
Finish: Mildly spicy, stays sweet until the ending.

Notes

Spirits of French Lick, West Baden, Indiana. A blend of sourced 7-year-old and French Lick's 2-year-old wheated bourbons. Some caramel malt used, too.

My Score

Notes: ______________________________

____

____

Appearance (1-5) ____ Nose (1-25) ____
Taste (1-25) ____ Finish (1-25) ____
Complexity (1-10) ____ Overall Impression (1-10) ____
SCORE ____

Whiskey Acres

Proof:	87
Age:	1 year
Type:	Not Straight
Style:	Farm Distilled
Mash Bill:	Not available
Color:	Straw
Price:	$$

Mike

Nose: Very young. Bread dough and candied fruit.
Taste: Corn bread dough and vanilla with some pepper.
Finish: Short and sweet with vanilla and spice.

Susan

Nose: Corn and oak with some leather and honey.
Taste: Jalapeno cornbread with some fruit.
Finish: Short and sweet with warmth at the end.

Notes

Made by Whiskey Acres Distilling Co., DeKalb, Illinois. Farm distillery that produces spirits from its own grain. They are experimenting with different corn varieties.

My Score

Notes: ______________________________

____

____

Appearance (1-5) ____ Nose (1-25) ____
Taste (1-25) ____ Finish (1-25) ____
Complexity (1-10) ____ Overall Impression (1-10) ____

SCORE ____

Whiskey Row

Proof: 88
Age: NAS
Type: Not Straight
Style: Small Batch
Mash Bill: Not available
Color: Light Straw
Price: $$$

Mike

Nose: Cherries and vanilla with a hint of oak.
Taste: Vanilla and cherries with some baking spice.
Finish: Starts fruity. Balance out with some oak.

Susan

Nose: Corn, light caramel, some spice.
Taste: Vanilla, cherry juice, some cinnamon.
Finish: Medium long and smooth.

Notes

Bourbon named for a section of Main Street in Louisville that historically was (and is once again) home to bourbon offices and distilleries.

My Score

Notes: ____________________________________

____

____

Appearance (1-5) ____ Nose (1-25) ____
Taste (1-25) ____ Finish (1-25) ____
Complexity (1-10) ____ Overall Impression (1-10) ____

SCORE ____

Whiskey Row Bottled-in-Bond

Proof: 100
Age: 4 years
Type: Straight
Style: Bonded, Four Grain
Mash Bill: Corn, rye, wheat, malted barley
Color: Amber
Price: $$$

Mike

Nose: Vanilla, peaches, apricots, oak, and spice.
Taste: Vanilla, apricots, baking spices, oak, a hint of leather.
Finish: Long with oak, spice, lingering fruit, and vanilla sweetness.

Susan

Nose: Grain forward and herbal with some light fruit and spice.
Taste: Dry popcorn, vanilla, some baking spice, and leather.
Finish: Smooth with light oak tannins.

Notes

Kentucky Artisan Distillery, Crestwood, Kentucky. Proportions of mash bill not disclosed.

My Score

Notes: ______________________________

____

____

Appearance (1-5) ____ Nose (1-25) ____
Taste (1-25) ____ Finish (1-25) ____
Complexity (1-10) ____ Overall Impression (1-10) ____
SCORE ____

Whiskey Sister

Proof: 80
Age: NAS
Type: Not Straight
Style: Small Batch, Craft
Mash Bill: Not available
Color: Light Straw
Price: $$

Mike

Nose: Very light. Corn and vanilla.
Taste: Sweet corn and vanilla with some honey apple.
Finish: Starts sweet and dries to oak and pepper ending.

Susan

Nose: Very light. Peaches and not much else.
Taste: Peaches on the entrance, then a mouthful of black pepper. Too hot for 80 proof.
Finish: Peaches, again, which linger to the spice.

Notes

Whiskey Sister, Fairfield, California. Bottled by Frank-Lin Distillers, Fairfield, California. Bourbon sourced from Kentucky and bottled in California.

My Score

Notes: ____________________________________

____

____

Appearance (1-5) ____ Nose (1-25) ____
Taste (1-25) ____ Finish (1-25) ____
Complexity (1-10) ____ Overall Impression (1-10) ____

SCORE ____

Widow Jane

Proof: 91
Age: 8 years
Type: Straight
Style: Single Barrel, Craft
Mash Bill: Not available
Color: Straw
Price: $$

Mike

Nose: Blackstrap molasses, cherries, oak.
Taste: Sweet caramel and pitted fruit – cherries and apricots.
Finish: Long and sweet with a hint of oak as it lingers.

Susan

Nose: Spicy corn, cherries. oak.
Taste: Spicy caramel with vanilla, pecan, ripe cherries.
Finish: Long, warm, ending in mild pepper spice.

Notes

Cacao Prieto Distillery, Brooklyn, New York. Also produces chocolate from the beans. Water comes from the limestone caves of the Widow Jane Mine in Rosendale, New York.

My Score

Notes: ______________________________

____

____

Appearance (1-5) ____ Nose (1-25) ____
Taste (1-25) ____ Finish (1-25) ____
Complexity (1-10) ____ Overall Impression (1-10) ____
SCORE ____

Widow Jane 10-Year-Old

Proof:	91
Age:	10 years
Type:	Straight
Style:	Non-Chill Filtered, Small Batch
Mash Bill:	Not available
Color:	Dark Straw
Price:	$$$

Mike

Nose: Cherry cotton candy with a hint of oak.
Taste: Thin mouthfeel. Cherries, vanilla, baking spices, and oak.
Finish: Long and dry with oak and spice.

Susan

Nose: Sweet corn, peaches, pistachios.
Taste: Immediate brown sugar and vanilla yield to more herbal notes and some oak.
Finish: Long and dry.

Notes

The Widow Jane small batch consists of 5 barrels.

My Score

Notes: ______________________________

____

____

Appearance (1-5)	____	Nose (1-25)	____
Taste (1-25)	____	Finish (1-25)	____
Complexity (1-10)	____	Overall Impression (1-10)	____

SCORE ____

Widow Jane Decadence

Proof:	91
Age:	NAS
Type:	Straight
Style:	Finished
Mash Bill:	Not available
Color:	Amber
Price:	$$$

Mike

Nose: Maple syrup, pecans, a hint of orange zest, and oak.

Taste: Maple syrup, pecans, oranges, cardamom, cloves, and white pepper with some oak.

Finish: Oak and spice with lingering sweetness from the maple syrup.

Susan

Nose: Toffee dipped in maple syrup, some light baking spice.

Taste: Rich toffee, sweet nuts, maple syrup. (Maple does not dominate, though.)

Finish: Maple lingers before drying to oaky caramel.

Notes

Aged 10 years before being finished in ex-maple syrup barrels.

My Score

Notes: ____

____

____

Appearance (1-5) ____ Nose (1-25) ____

Taste (1-25) ____ Finish (1-25) ____

Complexity (1-10) ____ Overall Impression (1-10) ____

SCORE ____

Widow Jane Lucky Thirteen

Proof: 93
Age: 13 years
Type: Straight
Style: Blend, Non-Chill Filtered
Mash Bill: Not available
Color: Amber
Price: $$$$

Mike

Nose: Rich caramel, apples, lots of oak, hint of spice.
Taste: Caramel and apples with a little pepper and oak.
Finish: Long with oak and pepper.

Susan

Nose: Very fruit forward with raisins, citrus peel, and dates imbedded in toffee.
Taste: Matched the nose with some addition of cinnamon and oak.
Finish: Long and oaky, but balanced with fruit and spice.

Notes

Rich and complex. Sourced from Kentucky, Tennessee, and Indiana.

My Score

Notes: ______________________________

____

____

Appearance (1-5) ____ Nose (1-25) ____
Taste (1-25) ____ Finish (1-25) ____
Complexity (1-10) ____ Overall Impression (1-10) ____

SCORE ____

Widow Jane The Vaults 2020 Release

Proof: 99
Age: 15 years
Type: Straight
Style: Blend, Non-Chill Filtered
Mash Bill: Not available
Color: Light Amber
Price: $$$$

Mike

Nose: Very brandy-like. Raisins and prunes with hints of citrus and oak.
Taste: Brandy-like richness. Raisins, berries, citrus zest, cardamom, white pepper, and oak.
Finish: Long with oak and spice and a coconut note.

Susan

Nose: Cola, prunes, dates, dark brown sugar, and orange peel.
Taste: Very savory. Dried dark fruit dusted in cocoa with cinnamon sticks.
Finish: Long and spicy ending in nutty oak.

Notes

Each annual release is different. This one was a blend of whiskeys 15 and 17 year bourbons sourced from Tennessee and Indiana. Complex.

My Score

Notes: ______________________________

____

____

Appearance (1-5) ____ Nose (1-25) ____
Taste (1-25) ____ Finish (1-25) ____
Complexity (1-10) ____ Overall Impression (1-10) ____

SCORE ____

Wiggly Bridge Small Barrel Bourbon

Proof: 96
Age: 4 years
Type: Straight
Style: Traditional
Mash Bill: 57% corn, 38% rye, 5% malted barley
Color: Light Straw
Price: $$$

Mike

Nose: Yeasty corn bread dough with a hint of oak.
Taste: Vanilla and corn with a hint of pepper.
Finish: Lots of corn with a hint of oak.

Susan

Nose: Cereal grain, tobacco, a bit of cinnamon, and fruit.
Taste: Vanilla, pear, and cinnamon.
Finish: Cereal note returns with touches of pepper and oak.

Notes

Wiggly Bridge Distillery, York, Maine. There's a bottled-in-bond expression, too.

My Score

Notes: ______________________________________

____

____

Appearance (1-5) ____ Nose (1-25) ____
Taste (1-25) ____ Finish (1-25) ____
Complexity (1-10) ____ Overall Impression (1-10) ____
SCORE ____

Wildcatter Barrel Select

Proof:	90
Age:	8 years
Type:	Straight
Style:	Traditional
Mash Bill:	Not available
Color:	Straw
Price:	$$

Mike

Nose: Caramel and pears with a hint of oak.
Taste: A bit thin. Caramel and pecans with baking spice.
Finish: Dry with oak and spice.

Susan

Nose: Corn husk, some caramel and dates.
Taste: Burnt caramel and dark fruit.
Finish: Dry and peppery.

Notes

Sourced from Kentucky and bottled by Elite Brands, Fairfield, California.

My Score

Notes: ____________________________________

____

____

Appearance (1-5) ____ Nose (1-25) ____
Taste (1-25) ____ Finish (1-25) ____
Complexity (1-10) ____ Overall Impression (1-10) ____

SCORE ____

William Dalton

Proof: 100
Age: 4 years
Type: Straight
Style: Wheated, Bonded
Mash Bill: 70% corn, 20% wheat, 10% malted barley
Color: Dark Straw
Price: $$

Mike

Nose: Caramel and milk chocolate with some pears, ripe apple, and oak.
Taste: Caramel and pears with cinnamon, nutmeg, and sweet oak.
Finish: Oak and spice with a hint of chocolate.

Susan

Nose: Vanilla pudding with sugar crystals, light pear, touch of cinnamon.
Taste: Vanilla cupcake with berries and white frosting.
Finish: Sweetness dries quickly to peppery oak.

Notes

As with all of the distllery's whiskeys, this is mostly available in Indiana and Kentucky.

My Score

Notes: ______________________________

____

____

Appearance (1-5) ____ Nose (1-25) ____
Taste (1-25) ____ Finish (1-25) ____
Complexity (1-10) ____ Overall Impression (1-10) ____
SCORE ____

Winchester

Proof: 90
Age: At least 2 years
Type: Straight
Style: Small Batch
Mash Bill: Not available
Color: Straw
Price: $$

Mike

Nose: Vanilla, oak, and corn in that order.
Taste: Very light. Vanilla and peaches with corn and oak.
Finish: Short and sweet. Just a hint of oak.

Susan

Nose: Honeyed vanilla and caramel.
Taste: Sweet corn, light caramel, faint fruit.
Finish: Medium long. Dry, but no burn.

Notes

Made by TerrePURE Spirits, North Charleston, South Carolina. A second distillery with the same owners started producing at the former Charles Medley facility in Owensboro, Kentucky in 2016. Both use a special process involving ultrasonic waves, heat, and oxygen to accelerate aging.

My Score

Notes: ______________________________

____

____

Appearance (1-5) ____ Nose (1-25) ____
Taste (1-25) ____ Finish (1-25) ____
Complexity (1-10) ____ Overall Impression (1-10) ____

SCORE ____

The Wiseman

Proof: 90.8
Age: NAS
Type: Straight
Style: Traditional
Mash Bill: Not available
Color: Straw
Price: $$

Mike

Nose: Very light. Corn, vanilla, and a hint of fruit.
Taste: Light. Corn, vanilla, cherries, and white pepper.
Finish: Medium long with oak and pepper.

Susan

Nose: Green apples, brown sugar, a suggestion of licorice.
Taste: Peaches, with brown sugar and cinnamon.
Finish: Medium long with sweet oak and pepper.

Notes

Owl, Jr.? Sourced from an unknown Kentucky distillery by the Kentucky Owl brand.

My Score

Notes: ______________________________

____

____

Appearance (1-5) ____ Nose (1-25) ____
Taste (1-25) ____ Finish (1-25) ____
Complexity (1-10) ____ Overall Impression (1-10) ____
SCORE ____

Wollersheim Bottled-in-Bond

Proof: 100
Age: 4 years
Type: Straight
Style: Bonded, Non-Chill Filtered
Mash Bill: White corn, rye, malted barley
Color: Straw
Price: $$$

Mike

Nose: Very fruity with apricots, plums, and raisins. Also, vanilla and oak.
Taste: Caramel, apricots, wild berries, cardamom, and oak.
Finish: Long with oak and pepper spice.

Susan

Nose: Caramel, mint, some raisins, and a little chocolate.
Taste: Dates and raisins with some dark cherries, caramel, oak, pepper, tobacco, and new leather.
Finish: Dries to long, lingering oak.

Notes

Wollersheim Winery & Distillery, Prairie Du Sac, Wisconsin.

My Score

Notes: ______________________________

____

____

Appearance (1-5) ____ Nose (1-25) ____
Taste (1-25) ____ Finish (1-25) ____
Complexity (1-10) ____ Overall Impression (1-10) ____
SCORE ____

Woodinville 8-Year-Old

Proof: 100
Age: 8 years
Type: Straight
Style: Limited Release
Mash Bill: 72% corn, 22% rye, 6% malted barley
Color: Light Amber
Price: $$$$

Mike

Nose: Caramel, pears, baking spices, oak.
Taste: Caramel, honey, pears, cinnamon, allspice, oak.
Finish: Long with oak and cinnamon.

Susan

Nose: Apples and strawberries with vanilla and nutmeg.
Taste: Caramel, cinnamon toast, apples, dark cherry, light oak.
Finish: Long, ending in oak and spice.

Notes

Exhibits depth and complexity. Bottle #515. Woodinville Distillery, Woodinville, Washington.

My Score

Notes: ______________________________________

____

____

Appearance (1-5) ____ Nose (1-25) ____
Taste (1-25) ____ Finish (1-25) ____
Complexity (1-10) ____ Overall Impression (1-10) ____

SCORE ____

Woodinville Port Finished Bourbon

Proof: 90
Age: NAS
Type: Straight
Style: Pot Distilled, Finished
Mash Bill: 72% corn, 22% rye, 6% malted barley
Color: Amber
Price: $$

Mike

Nose: Very fruity with raisins, plums, vanilla, and oak.
Taste: Raisins and berries with vanilla, cinnamon, oak, and a hint of dark chocolate.
Finish: Long and dry with oak, cinnamon, and hint of chocolate.

Susan

Nose: Distinct, but not overpowering, wine note with vanilla and milk chocolate.
Taste: Chocolate, cherries, and caramel.
Finish: Dries to spicy tannins and chocolate.

Notes

Website states this is 5 years old with additional 6 months in the port barrels.

My Score

Notes: ______________________________

____

____

Appearance (1-5) ____ Nose (1-25) ____
Taste (1-25) ____ Finish (1-25) ____
Complexity (1-10) ____ Overall Impression (1-10) ____

SCORE ____

Woodinville Straight Bourbon

Proof: 118.1
Age: NAS
Type: Straight
Style: Pot Distilled
Mash Bill: 72% corn, 22% rye, 6% malted barley
Color: Light Amber
Price: $$

Mike

Nose: Corn, vanilla, a little fruit, oak.
Taste: Corn, vanilla, pear, cinnamon, oak.
Finish: Long with oak, pepper, and a bit of chocolate.

Susan

Nose: Roasted corn, brown sugar, some berries.
Taste: Rich caramel, cinnamon, apples, and a bit of chocolate.
Finish: Long and dry with oak and cinnamon.

Notes

Woodinville Whiskey Company. Woodinville Washington. Nicely complex. All grains sourced from one farm in Washington.

My Score

Notes: ____

____

____

Appearance (1-5) ____ Nose (1-25) ____
Taste (1-25) ____ Finish (1-25) ____
Complexity (1-10) ____ Overall Impression (1-10) ____

SCORE ____

Woodstone Creek

Proof: 94
Age: NAS*
Type: Straight
Style: Single Barrel, Craft
Mash Bill: Five grains**
Color: Dark Straw
Price: $$$$

Mike

Nose: Very malty – corn and barley malt. Caramel and vanilla with hint of sweet spices and tobacco.
Taste: Sweet with malt and corn. A bit of pepper spice and oak. Touch of hickory nuts and hazelnuts.
Finish: Long and spicy turning to nice, dry oak.

Susan

Nose: Light caramel, popcorn, light pepper, tobacco.
Taste: Vanilla and caramel. Not-quite-ripe apricots and toasted pecans.
Finish: Long and dry. Ends with black pepper.

Notes

Woodstone Creek Distillery, Cincinnati, Ohio. Distiller told us it is 5 years old.* No proportions given, but made with white corn, yellow corn, malted rye, malted wheat, and malted barley.** Made in a pot still. Only 2-3 barrels made per year and price weighs in at about $100/bottle.

My Score

Notes: ______________________________
____
____

Appearance (1-5) ____ Nose (1-25) ____
Taste (1-25) ____ Finish (1-25) ____
Complexity (1-10) ____ Overall Impression (1-10) ____

SCORE ____

Wyoming Whiskey

Proof: 88
Age: 3 years
Type: Straight
Style: Small Batch, Craft, Wheated
Mash Bill: 68% corn, 20% wheat, 12% malted barley
Color: Straw
Price: $$

Mike

Nose: Sweet honey and ripe apple. Vanilla and corn.
Taste: Sweet honey and corn with some spice. Allspice and apples.
Finish: Long and sweet with a bit of spice.

Susan

Nose: Corn an oak and a little clove.
Taste: Vanilla and apples.
Finish: Long and warm. Some spice.

Notes

Wyoming Whiskey, Kirby, Wyoming. Batch 14, Bottled June 13, 2013. This would be an even better bourbon at 4 years old. Gets better as it breathes.

My Score

Notes: ______________________________

____

____

Appearance (1-5) ____ Nose (1-25) ____
Taste (1-25) ____ Finish (1-25) ____
Complexity (1-10) ____ Overall Impression (1-10) ____

SCORE ____

Wyoming Whiskey Double Cask

Proof: 100
Age: NAS
Type: Straight
Style: Wheated, Finished
Mash Bill: 68% corn, 20% wheat, 12% malted barley
Color: Light Amber
Price: $$$

Mike

Nose: Sherry fruit – raisins, cherries – with some almond and oak. Vanilla and spice in the background.
Taste: Very brandy-like with cherries, berries, vanilla, and baking spices.
Finish: Medium long and fruity with only a hint of oak.

Susan

Nose: Baked apples and caramel, with a little new leather.
Taste: Rich vanilla and fruit with some baking spices.
Finish: Stays sweet for a long time before ending in spice.

Notes

Finished in sherry casks after 5 years in original barrels.

My Score

Notes: ____________________________________
____
____

Appearance (1-5) ____ Nose (1-25) ____
Taste (1-25) ____ Finish (1-25) ____
Complexity (1-10) ____ Overall Impression (1-10) ____
SCORE ____

Wyoming Whiskey Single Barrel

Proof: 96 (and up)
Age: 5 years
Type: Straight
Style: Wheated, Cask Strength
Mash Bill: 68% corn, 20% wheat, 12% malted barley
Color: Light Amber
Price: $$$

Mike

Nose: Vanilla, cherries, oak, baking spices with some fine leather.
Taste: Vanilla and cherries with lots of nutmeg and cinnamon.
Finish: Long and dry with oak tannins and a hint of chocolate.

Susan

Nose: Caramel, dark fruit, some cinnamon.
Taste: Caramel, vanilla, loads of baking spice, some cashews and cherries.
Finish: Oaky and spicy.

Notes

Unusually spicy for a wheated bourbon.

My Score

Notes: ____________________________________

____

____

Appearance (1-5) ____ Nose (1-25) ____
Taste (1-25) ____ Finish (1-25) ____
Complexity (1-10) ____ Overall Impression (1-10) ____

SCORE ____

Yellow Rose Double Barrel

Proof: 86
Age: NAS
Type: Straight
Style: Special Finish
Mash Bill: Not available
Color: Dark Straw
Price: $$

Mike

Nose: Caramel and figs. Hint of spice.
Taste: Caramel and fruit. Hint of butterscotch.
Finish: Long and fruity with butterscotch and oak.

Susan

Nose: Light caramel, dark fruit, oak.
Taste: Caramel and oak with touch of black pepper and tobacco.
Finish: Short and warm. No burn.

Notes

Wine barrel finished. This was bottle #618 from Batch #1. From Yellow Rose Distilling, Houston, Texas.

My Score

Notes: ______________________________

____

____

Appearance (1-5) ____ Nose (1-25) ____
Taste (1-25) ____ Finish (1-25) ____
Complexity (1-10) ____ Overall Impression (1-10) ____

SCORE ____

Yellow Rose Outlaw

Proof: 92
Age: 6 months
Type: Not Straight
Style: Small Batch
Mash Bill: Not available
Color: Light Amber
Price: $$$

Mike

Nose: Corn and vanilla. Hint of oak and pecans.
Taste: Corn and vanilla with some pecans and oak.
Finish: Dry with pecans and oak.

Susan

Nose: Oak, nuts, some sweet spice.
Taste: Oak and corn, plus pecans.
Finish: Very dry/woody.

Notes

From Yellow Rose Distilling, Houston, Texas. Bottle #467 from batch #27. Aged in 25-gallon casks.

My Score

Notes: ______________________________

____

____

Appearance (1-5) ____ Nose (1-25) ____
Taste (1-25) ____ Finish (1-25) ____
Complexity (1-10) ____ Overall Impression (1-10) ____

SCORE ____

Yellowstone Limited Release 2021

Proof: 101
Age: NAS
Type: Straight
Style: Blend, Finished
Mash Bill: 51% corn, 45% rye, 4% malted barley
Color: Light Amber
Price: $$$$

Mike

Nose: Caramel and raisins with prunes, spice, and lots of oak.
Taste: Caramel and raisins with some dried berries, baking spice, and oak.
Finish: Long with oak and spice and some lingering fruit.

Susan

Nose: Dark fruit, toffee, butterscotch.
Taste: Fruit medley, sweet baking spices, and pepper. Warm but not hot.
Finish: Flavors fade to pepper. Add water for more fruit.

Notes

Limestone Branch Distillery, Lebanon, Kentucky. Blend of 7 and 15 year-old bourbons. Each annual edition is different. This one finished in Amarone casks. Other distillery expressions include Limestone Branch Single Barrel Experimental edition.

My Score

Notes: ______________________________

____

____

Appearance (1-5) ____ Nose (1-25) ____
Taste (1-25) ____ Finish (1-25) ____
Complexity (1-10) ____ Overall Impression (1-10) ____

SCORE ____

Yellowstone Select

Proof: 93
Age: NAS
Type: Straight
Style: Small Batch
Mash Bill: Not available
Color: Straw
Price: $$$

Mike

Nose: Corn and vanilla with a bit of tobacco. A bit yeasty.
Taste: Corn and vanilla with pears, oak, and spice.
Finish: Very nice. Starts dry with oak and gets spicy.

Susan

Nose: Mint, caramel, some orange zest.
Taste: Vanilla and caramel, some chocolate and pears. Rich mouthfeel.
Finish: Warm and smooth ending with a nice milk chocolate note.

Notes

Sourced from Kentucky and bottled by Limestone Branch Distillery, Lebanon, Kentucky. It is owned and distributed by Luxco, St. Louis, Missouri.

My Score

Notes: ______________________________

____

____

Appearance (1-5) ____ Nose (1-25) ____
Taste (1-25) ____ Finish (1-25) ____
Complexity (1-10) ____ Overall Impression (1-10) ____

SCORE ____

Zachory Boone

Proof: 80
Age: NAS
Type: Straight
Style: Traditional
Mash Bill: Not available
Color: Light Straw
Price: $

Mike

Nose: Very, very light nose. Corn and vanilla with a hint of toasted marshmallow.
Taste: Sweet with vanilla and light fruit – bananas and pears.
Finish: Medium long and spicy with pepper spice.

Susan

Nose: Very light. Some oak, corn, and vanilla.
Taste: Ripe apples, vanilla, light pepper.
Finish: Short and oaky.

Notes

Bottled by Brookdale Ltd., Mira Lona, California.

My Score

Notes: ____________________________________

____

____

Appearance (1-5) ____ Nose (1-25) ____
Taste (1-25) ____ Finish (1-25) ____
Complexity (1-10) ____ Overall Impression (1-10) ____
SCORE ____

Other Bourbons Additional Releases

Name: ______________________________

Distillery: ______________________________

Proof: ____________ **Mash Bill:** ____________
Age: ____________ ____________
Type: ____________ **Color:** ____________
Style: ____________ **Price:** ____________

My Score

Notes: ______________________________

____

____

Appearance (1-5) ____ Nose (1-25) ____
Taste (1-25) ____ Finish (1-25) ____
Complexity (1-10) ____ Overall Impression (1-10) ____
SCORE ____

Name: ______________________________

Distillery: ______________________________

Proof: ____________ **Mash Bill:** ____________
Age: ____________ ____________
Type: ____________ **Color:** ____________
Style: ____________ **Price:** ____________

My Score

Notes: ______________________________

____

____

Appearance (1-5) ____ Nose (1-25) ____
Taste (1-25) ____ Finish (1-25) ____
Complexity (1-10) ____ Overall Impression (1-10) ____
SCORE ____

Other Bourbons Additional Releases

Name: ______________________

Distillery: ______________________

Proof: __________ **Mash Bill:** __________
Age: __________ __________
Type: __________ **Color:** __________
Style: __________ **Price:** __________

My Score

Notes: ______________________

____

____

Appearance (1-5) ____ Nose (1-25) ____
Taste (1-25) ____ Finish (1-25) ____
Complexity (1-10) ____ Overall Impression (1-10) ____
SCORE ____

Name: ______________________

Distillery: ______________________

Proof: __________ **Mash Bill:** __________
Age: __________ __________
Type: __________ **Color:** __________
Style: __________ **Price:** __________

My Score

Notes: ______________________

____

____

Appearance (1-5) ____ Nose (1-25) ____
Taste (1-25) ____ Finish (1-25) ____
Complexity (1-10) ____ Overall Impression (1-10) ____
SCORE ____

Other Bourbons Additional Releases

Name: ______________________________

Distillery: ______________________________

Proof: ____________ **Mash Bill:** ____________
Age: ____________ ____________
Type: ____________ **Color:** ____________
Style: ____________ **Price:** ____________

My Score

Notes: ______________________________
____
____

Appearance (1-5) ___ Nose (1-25) ___
Taste (1-25) ___ Finish (1-25) ___
Complexity (1-10) ___ Overall Impression (1-10) ___

SCORE ___

Name: ______________________________

Distillery: ______________________________

Proof: ____________ **Mash Bill:** ____________
Age: ____________ ____________
Type: ____________ **Color:** ____________
Style: ____________ **Price:** ____________

My Score

Notes: ______________________________
____
____

Appearance (1-5) ___ Nose (1-25) ___
Taste (1-25) ___ Finish (1-25) ___
Complexity (1-10) ___ Overall Impression (1-10) ___

SCORE ___

Other Bourbons Additional Releases

Name: ______________________________

Distillery: ______________________________

Proof: ____________ **Mash Bill:** ____________
Age: ____________ ____________
Type: ____________ **Color:** ____________
Style: ____________ **Price:** ____________

My Score

Notes: ______________________________

____

____

Appearance (1-5) ____ Nose (1-25) ____
Taste (1-25) ____ Finish (1-25) ____
Complexity (1-10) ____ Overall Impression (1-10) ____

SCORE ____

Name: ______________________________

Distillery: ______________________________

Proof: ____________ **Mash Bill:** ____________
Age: ____________ ____________
Type: ____________ **Color:** ____________
Style: ____________ **Price:** ____________

My Score

Notes: ______________________________

____

____

Appearance (1-5) ____ Nose (1-25) ____
Taste (1-25) ____ Finish (1-25) ____
Complexity (1-10) ____ Overall Impression (1-10) ____

SCORE ____

Other Bourbons Additional Releases

Name: ______________________________

Distillery: ______________________________

Proof: ____________ **Mash Bill:** ____________
Age: ____________ ____________
Type: ____________ **Color:** ____________
Style: ____________ **Price:** ____________

My Score

Notes: ______________________________

____

____

Appearance (1-5) ___ Nose (1-25) ___
Taste (1-25) ___ Finish (1-25) ___
Complexity (1-10) ___ Overall Impression (1-10) ___
SCORE ___

Name: ______________________________

Distillery: ______________________________

Proof: ____________ **Mash Bill:** ____________
Age: ____________ ____________
Type: ____________ **Color:** ____________
Style: ____________ **Price:** ____________

My Score

Notes: ______________________________

____

____

Appearance (1-5) ___ Nose (1-25) ___
Taste (1-25) ___ Finish (1-25) ___
Complexity (1-10) ___ Overall Impression (1-10) ___
SCORE ___

Other Bourbons Additional Releases

Name: ______________________________

Distillery: ______________________________

Proof: ____________ **Mash Bill:** ____________
Age: ____________ ____________
Type: ____________ **Color:** ____________
Style: ____________ **Price:** ____________

My Score

Notes: ______________________________

____

____

Appearance (1-5) ____ Nose (1-25) ____
Taste (1-25) ____ Finish (1-25) ____
Complexity (1-10) ____ Overall Impression (1-10) ____

SCORE ____

Name: ______________________________

Distillery: ______________________________

Proof: ____________ **Mash Bill:** ____________
Age: ____________ ____________
Type: ____________ **Color:** ____________
Style: ____________ **Price:** ____________

My Score

Notes: ______________________________

____

____

Appearance (1-5) ____ Nose (1-25) ____
Taste (1-25) ____ Finish (1-25) ____
Complexity (1-10) ____ Overall Impression (1-10) ____

SCORE ____

Acknowledgments

Many people helped in the preparation of the book. We received samples from several distilleries, as well as bottle photographs. We also appreciated information about ages and non-propriety mash bills. Thanks go to Ginger Flowers and Linda Hayes (Beam Suntory), Tracy Frederick, Phil Lynch, and Chris Morris (Brown-Forman), Amy Preske (Buffalo Trace/Barton 1792), Ellen King, Karen Kushner, and Al Young (Four Roses) Josh Hafer, Kelly Hubbuch, Larry Kass, and Niki King (Heaven Hill), Shelby Caret (High West Whiskey), Julian Van Winkle, III (Pappy Van Winkle), Tim Quinlan (River City Distributors), John Foster (Smooth Ambler) and Katrina Egbert and Rick Robinson (Wild Turkey).

And many thanks to the scores of craft distillers and independent bottlers who sent us photos and samples, not to mention our friend Julie Bartlett, who brought bourbons from her trip to Oregon.

The authors also tasted bourbons in several venues around Louisville and we enjoyed meals afterwards. Thank you Jason Brauner and John Morrison (Bourbons Bistro), Bonnie Eisert and Brian Sur (Charr'd Bourbon Kitchen and Lounge at the Louisville Marriott East) Anderson Grissom (The Dish), and Larry Rice (The Silver Dollar). Chris Zaborowski at Westport Whiskey & Wine gave us access to bourbons in the tasting room that we did not have.

Many thanks also go to the bourbon lovers who occasionally joined us to taste and comment on the bourbons at hand (after we had made notes, of course). They were Dan Gardner, Joanna Goldstein, Sean Higgins, Rosemary Miller, and Stacey Yates. Cheers!

The Authors

Susan Reigler

Bourbon Authority and Author

Award-winning writer Susan Reigler, author of *Kentucky Bourbon Country: The Essential Travel Guide, The Complete Guide to Kentucky State Parks, More Kentucky Bourbon Cocktails*, and co-author of *The Kentucky Bourbon Cocktail Book*, was born in Louisville the year Swaps won the Kentucky Derby. From 1992 to 2007, Reigler was a restaurant critic, beverage columnist and travel writer for the Louisville *Courier-Journal.* She is a Certified Executive Bourbon Steward by the Society of the Stave & Thief and a member of the Order of the Writ.

With her cat, E.T.

Reigler served as the President of the Board of Directors of the Bourbon Women Association from 2015 to 2017. She has presented workshops at the Kentucky Bourbon Festival, Bourbon & Beyond (Louisville), and bourbon festivals in Chicago, New Orleans, Oxford, MS, Smithfield, VA and others. She has lead bourbon tastings from Seattle to Savannah as well tastings to benefit non-profit organizations including Locust Grove, The Falls of the Ohio Foundation, and Shaker Village at Pleasant Hill. She has been a judge for international spirits competitions and bourbon cocktail contests and has helped restaurants and retailers select barrels from Four Roses, Woodford Reserve, Knob Creek, and more, for private bourbon bottlings.

Reigler currently writes for *American Whiskey Magazine* and *Barleycorndrinks.com*. She holds a bachelor's degree in music from Indiana University and a master's degree in zoology from the University of Oxford, which she attended as a Humphrey Scholar.

Michael R. Veach

Bourbon Historian and Author
The Filson Historical Society

Mike Veach is one of the foremost authorities on all things Bourbon. He spent over 15 years with The Filson Historical Society serving in the role of historian. Veach possesses an unsurpassed wealth of knowledge and perspective on an industry that has Kentucky at the forefront of the Bourbon Boom occurring across the United States and abroad.

Veach is also a recognized author whose articles have been featured in publications such as *The Bourbon Review* and *The Louisville Encyclopedia*. He wrote the foreword for the release of Gerald Carson's *Social History of Bourbon in 2010*, and in 2013 published his own now-widely acclaimed book *Kentucky Bourbon Whiskey: An American Heritage*.

Prior to his long tenure at The Filson, Veach spent 5 years as the North American Archivist for United Distillers. Since then, he has also provided research and consulting services for such institutions as the Oscar Getz Museum of Whiskey History, Brown-Forman, Buffalo Trace, Four Roses, and Heaven Hill Distillery.

In 2006, Mike Veach was publicly recognized for his knowledge, activities, and accomplishments surrounding bourbon by being inducted into The Kentucky Bourbon Hall of Fame. In 2019 he was inducted into the Order of the Writ. He has provided pro-bono services to many major non-profit entities such as The Black Acre Foundation, The Farnsley-Kaufman House, The Farnsley-Moreman Landing, Boy Scouts of America, Girls. Inc in Owensboro and many other organizations including The Filson Historical Society.

Veach holds a BA and MA in History from the University of Louisville. He travels across the country researching, lecturing, and hosting special seminars designed to spread the knowledge of Kentucky Whiskey products and enhance the appreciation of Bourbon. Mr. Veach resides in his hometown of Louisville, KY in the heart of Bourbon Country where he writes the blog *bourbonveach.com.*

INDICES

Alphabetical

Proof

Special Style

Barrel Proof/Cask Strength

(No water added before bottling)

Bottled-in-Bond/Bonded

(Bourbon that is at least 4 years old, bottled at 100 proof, and produced by one distiller at one distillery in one season)

Extra Aged
(10 years and more)

Farm Distilled

Five Grain

Four Grain

Non-Chill Filtered/Unchilled/Unfiltered

Single Barrel

(Bottled from a single barrel. Can yield ~ 60-150 bottles per barrel depending on how much evaporation took place during maturation.)

Six Grain

Special Finish and/or Secondary Barrel

Two Grain

Wheated

Price

Subject to change – MSRP, not secondary market price.

$$$ *($50 – $99.99)*

State

Where ultimately bottled and parent company located.
If sourced from a different state, that is noted in the description.

My Top 25 Bourbons

Name	Total Score	Page #
1.		
2.		
3.		
4.		
5.		
6.		
7.		
8.		
9.		
10.		
11.		
12.		
13.		
14.		
15.		
16.		
17.		
18.		
19.		
20.		
21.		
22.		
23.		
24.		
25.		